CONTENTS

That is the essence of science: ask an impertinent question, and you are on the way to a pertinent answer.

Jacob Bronowski

PROLOGUE

I t is 5.15 pm on 23 March 2003. In a brightly lit auditorium in Davis, California, Harvard cosmologist Lisa Randall is trying to give a talk about her research. The audience contains some of the greatest scientific minds on the planet, even some Nobel laureates, but no one is paying Randall any attention. Even she is having trouble concentrating. Her eyes flick repeatedly from her notes to the front row of the audience. There, on the far right of the auditorium, Stephen Hawking is being given his tea-time soup. It's quite a sight.

Earlier in the day Hawking gave a sparkling talk, crammed with witty asides and acerbic commentaries on the state of science. It was delivered via his speech synthesiser, with that hallmark monotony; Hawking is paralysed by motor neuron disease and simply cannot speak for himself. Eating is similarly problematic.

His nurses are trying their best to avoid a spectacle, but it is difficult. The spoon won't quite go into his mouth, and the soup dribbles down his chin. It is unquestionably distracting: not one of these fine minds has the capacity to ignore the goings-on in the front row and focus exclusively on Randall's talk. Discomfiting as this scenario is, there is an upside. Here, in this strange moment of their lofty, cerebral lives, it has become clear, just for a moment, that these scientists are very human beings.

The humanity of scientists – and what that really means – is what this book is about. For more than fifty years, scientists have been involved in a cover-up that is arguably one of the most successful of modern times. It has succeeded because even the scientists haven't understood what has been going on.

After the Second World War, science was given a makeover. It was turned into a brand – in the same way that Coca-Cola, Apple Computers, Disney and McDonald's are brands. The brand identity of science is reinforced with adjectives such as logical, responsible, trustworthy, predictable, dependable, gentlemanly, straight, boring, unexciting, objective, rational. Not in thrall to passions or emotion. A safe pair of hands. In summary: unhuman.

The creation and protection of this brand – the perpetuating of the myth of the rational, logical scientist who follows a clearly understood Scientific Method – has coloured everything in science. It affects the way it is done, the way we teach it, the way we fund it, its presentation in the media, the way its quality control structures – in particular, peer review – work (or don't work), the expectation we have of science's impact on society, and the way the public engages with science (and scientists with the public) and regards scientists' pronouncements as authoritative. We have been engaging with a caricature of science, not the real thing. But science is so vital to our future that it must now be set free from its branding. It is time to reveal science as the anarchic, creative, radical endeavour it has always been.

Science's domination of today's world belies the fact that it is a relative newcomer as a profession – perhaps one of the newest.

Before the Second World War, jobs in science were largely ivory tower affairs reserved for the few. However, the global conflict showed that scientists were capable of changing the fates of nations. During those difficult years, science provided governments and their armies with penicillin, radar and – of course – the atomic bomb, among myriad other innovations. Those in power quickly realised that science was a good investment: if there should be another war, then whoever had the best scientists would win. Physicists were 'the Merlins of the Cold War', as Michael Schrage has put it: 'their wizardry could tip the balance of the superpowers in the twinkling of a quark'.

What followed, according to historian Steven Shapin, was the 'professionalisation and routinisation of science as a remunerated job'. So, with the prospect of secure funding, steady jobs and even good pensions, scientists set about making themselves look worthy of the investment. The first task was to solve their image problem.

At the end of the Second World War, when this process began, scientists were mistrusted. Though their power was enticing to governments, it was also disturbing. 'The Stone Age may return on the gleaming wings of Science,' warned Winston Churchill, 'and what might now shower immeasurable material blessings upon mankind may even bring about its total destruction.'

Another of Churchill's pronouncements makes science's dilemma plain:

> It is arguable whether the human race have been gainers by the march of science beyond the steam engine. Electricity opens a field of infinite conveniences to ever greater numbers, but they may well have to pay dearly for them. But anyhow in my thought I stop short of the internal combustion engine which has made the world so much smaller.

Still more must we fear the consequences of entrusting a human race so little different from their predecessors of the so-called barbarous ages such awful agencies as the atomic bomb. Give me the horse.

The fear of science's power is almost palpable. Penicillin and radar had helped the Allies survive the conflict, but it was the scientists' cataclysmic unleashing of atomic energy that won it. And it was the scientific mind that produced the rockets that had rained down on London, causing such devastation and misery. Tales about the inhumanity of science were leaking out too: reports of scientists conducting gruesome and inhuman experiments in the German concentration camps, and of Japanese medical research on prisoners of war. Churchill would also have known of Allied scientists testing nerve gas and mustard gas on their own soldiers.

The scientists' first move was to dissipate the unease the public felt about science's power and sense of responsibility; science would henceforth serve the people. Science projected itself as responsible and safe: a careful, measured discipline involving sensible, level-headed people not given to dangerous passions. As the renowned biologist and broadcaster Jacob Bronowski put it just a few years after Hiroshima, the scientist became 'the monk of our age, timid, thwarted, anxious to be asked to help'.

It was a deliberate policy: whenever British scientists of the post-war era allowed television cameras into their laboratories, for example, the message was upbeat and optimistic, 'very much the image of science that the high-ups in the Royal Society wanted to put across', as Tim Boon, chief curator of London's Science Museum, has put it. Television drama, on the other hand, free from the influence of senior scientists, showed a much more distrustful attitude. 'You scientists,' rages a character in a 1960s drama, 'you kill half the world, and the other half can't live without you.'

Once the scientists' subservience was established, all they had to do was convince governments and the public that science had at its disposal a safe, efficient, controllable Method that, given enough resources, they could use to create a better world. It helped that science works so well. By 1957, 96 per cent of Americans said they agreed with the statement that 'science and technology are making our lives healthier, easier and more comfortable'.

The scientists too allowed themselves to be fooled by the cover-up. They became convinced that they were the heirs to a noble and dispassionate tradition, and that the brand values of science were carefully nurtured and passed down the scientific generations. According to the US Office of Technology Assessment, the average science professor trains around twenty PhD scientists. All are, almost unconsciously, taught to play by a set of rules that will perpetuate the myth of the responsible, level-headed, trustworthy scientist.

One of the few senior scientists to have dared to expose the spin was the British biologist and Nobel laureate Peter Medawar. Scientists, he admitted, 'actively misrepresent' themselves. The famed scientific routine of deductions based on experiments that were themselves based on logical hypotheses 'are simply the postures we choose to be seen in when the curtain goes up and the public sees us', Medawar said. 'The illusion is shattered if we ask what goes on behind the scenes.'

So what *does* go on behind the scenes? The most concise description was given by the Austrian-born physicist turned philosopher Paul Feyerabend. In 1975, Feyerabend published a book called *Against Method* in which he set out a shocking idea. When it comes to pushing at the frontiers of knowledge, there is only one rule, he said: Anything Goes. Science is anarchy.

Feyerabend was soon declared the 'worst enemy of science', and for good reason. His argument was deliberately provocative and mischievous, and he took it to the furthest extremes: witchcraft was just as valid a way of gathering knowledge, he once contended. But his point still stands. When we look behind the curtain, science is astonishing.

To make a breakthrough or to stay on top, scientists take drugs, they follow crazy dreams, they experiment on themselves and on one another, and occasionally they die in the process. They fight – sometimes physically, but mostly in intellectual battles. They try to entrap one another, standing in their colleagues' way to block progress and maintain the lead. They break all the rules of polite society, trampling on the sacred, showing a total disregard for authority. They commit fraud or deceive or manipulate others in order to get to the truth about how the world works. They conjure up seemingly ridiculous ideas, then fight tooth and nail to show that the ideas are not only far from ridiculous, but exactly how things really are. Some challenge the interests of government and business, occasionally sacrificing their reputations for the greater good. Science is peppered with successes that defy rational explanation, and failures that seem even more illogical. There are moments of euphoria and – just once in ten thousand working lifetimes – world-changing success.

This is not the 'wacky' science, the crazy things that happen on the fringes of research. This is the mainstream. These anarchies are behind many of the Nobel Prizes of the last few decades – the decades that have given us such powerful insights into what the universe is, how it works and how we fit into its schemes. It really does seem that, in science, anything goes.

And this is no modern phenomenon. Science has always been this way, because this is how it works. Isaac Newton, for instance, was cavalier with scientific truth, and cared nothing for the accepted rules of engagement. His writings contain passages that his biographers have declared to be 'nothing short of deliberate fraud'. He routinely made discoveries then kept them to himself, taunting his colleagues about his 'secret knowledge'.

Newton is known for humbly declaring that he had achieved his great breakthroughs by 'standing on the shoulders of giants'. Though this may be true in part, it is largely humbug. Newton was hardly humble, and it would be just as true to say that he achieved greatness by stamping on the shoulders of giants. When others, such as Robert Hooke and Gottfried Liebniz, made breakthroughs in fields he was also researching, Newton fought ferociously to deny them credit for their work. Though his reputation has been polished for centuries – he is the 'scientist's scientist' – Newton was not someone you would want to put in charge of science today; in later life he suffered episodes of madness and became obsessed with the Old Testament Book of Daniel, writing a commentary on it that he considered his greatest work. Hardly the model of scientific level-headedness.

Albert Einstein, who is widely considered to be the greatest scientist in history after Newton, provides another classic and shocking example of the reality behind scientific progress. Einstein relied on mystical insights – insights that his mathematics was not good enough to prove. His papers are riddled with errors and convenient omissions – though they were lazy fudges rather than, as with Newton, deliberate frauds. Einstein repeatedly failed to take account of known facts when formulating his ideas. He bristled at reviewers' criticisms of his papers. More than once he argued that any data found to be in conflict with his beautiful ideas should be ignored. He took credit for the $E = mc^2$ equation even though he wasn't the first to suggest it. Neither did he ever manage to prove it, despite eight published attempts: it was left to other, better mathematicians to set the world's most famous equation on the firm footing it has today.

History, they say, is written by the winners. Perhaps that's why

Galileo Galilei is also known as a hero, not a fraud. His *Dialogue Concerning the Two Chief Systems of the World*, banned for two centuries by the Catholic Church because it provides a bedrock for the heliocentric universe, is riddled with glaring errors. Though this monograph earned him a life sentence under house arrest, Galileo was no martyr to the truth: in many places, his science simply does not stand up. Given the man's obvious brilliance, historians now concede that his errors are an attempt at fraud resulting from obsession. Galileo was so convinced that the Earth moved round the Sun that he wasn't prepared to let the difficulties of making a watertight argument get in the way.

As we will see in the pages that follow, the tradition of scientific anarchy continues right up to the present day – though today's anarchies are much better concealed. But the purpose of this book is not just to present a string of entertaining anecdotes about scientific 'misbehaviour'. Its purpose is to show how scientists get the job done, and to argue that our misplaced expectations of science are preventing further discovery. This brand identity is not how science *really* is, and the disparity between the public image of science and the way breakthroughs are actually made matters more than most people realise or care to acknowledge. Scientists are starting to accept the straitjacket of the robot-researcher as if it were a standard-issue lab coat, necessary for the job. The fact is, you can't do good science in a straitjacket. This book is a call for more scientific anarchy, and for the creation of a culture in which it can thrive. After all, our future may depend on it.

On 20 November 2009, the world woke up to the 'climategate' scandal. Activists sceptical of scientists' claims about climate change had hacked into the email system of the University of

East Anglia. They managed to download a set of communications which, the activists claimed, showed that scientists had manipulated climate data to strengthen the case for global warming.

The ensuing investigation eventually cleared the scientists involved of any scientific misdemeanours, but there were serious official misgivings about some of the scientists' attitudes and obstructiveness towards those trying to get hold of their data. And the damage, it seemed, was done. In February 2010, a poll commissioned by the BBC showed that the number of adults who did not think global warming was happening had increased by 10 per cent since the previous November. This was 'very disappointing', Bob Watson, the UK's chief environmental scientist, told BBC News. 'Trust has been damaged,' German climate scientist Hans von Storch told the *Guardian* in July 2010. 'People now find it conceivable that scientists cheat and manipulate.'

The thing is, this doesn't actually explain the BBC poll results. Close inspection reveals that most people who had changed their views as a direct result of climategate had become *more* convinced of global warming, not less.

The downturn in public acceptance of climate change was most likely a consequence of a harsh British winter. A study carried out in March by Stanford University researchers revealed that any impact of climategate on public opinion had already disappeared. This was confirmed in June, when polls on both sides of the Atlantic showed that February's increase in climate scepticism had died away.

The only tangible outcome of climategate was positive. People who were unsure about whether to trust scientists got a glimpse of scientists being human – and thought that was OK. In fact it was more than OK, as the net allegiance change in the BBC poll shows. Contrary to everything scientists might have feared, exposing their irrationality, their humanity, even their craftiness and hot tempers, makes the public more receptive to the revelations of science, not less. People can not only *take* the truth about science, they actually *prefer* it.

It seems that scientists may have perpetrated one of the most misguided cover-ups in history. The trouble is, it will be painful to undo because it has served some scientists rather well.

The educated Western mind venerates science to the point of mysticism: its proponents are the new high priests. And scientists do little to discourage that reverence. In his 1951 book *The Common Sense of Science*, Bronowski went so far as to admit that scientists actively welcome it. Scientists 'have enjoyed acting the mysterious stranger, the powerful voice without emotion, the expert and the god,' he wrote. A famous example of this comes at the end of Hawking's extraordinary book, *A Brief History of Time*. He talks about the revelations we are seeking from science. Get to where we want to go, he says, and we will 'know the mind of God'.

The scientists with Hawking in the Davis auditorium are closer to knowing the mind of God than most. The meeting was convened to discuss the implications of a new set of results from one of NASA's orbiting telescopes: the Wilkinson Microwave Anisotropy Probe, or WMAP. WMAP is a satellite equipped with state-of-the-art instrumentation and backed up by thousands of researchers who use the world's biggest computers to dissect its data. But its function can be summed up pretty simply: it is a pair of cosmic bat ears.

Just as bats listen for echoes to tell them what is in their surroundings, WMAP listens for echoes – in the form of heat radiation – from the early universe to tell us what was there. We are blind to the first moments of creation because they took place too long ago. But we can still pick up the echoes, and these echoes are clear enough to give us insights into the beginning of everything.

They tell us, for instance, when and how the first atoms formed, and that in turn is enough to tell us when the first subatomic particles formed, and when the forces of nature first appeared, right back to an infinitesimal fraction of time after the Big Bang itself. Thanks to the WMAP probe, and other experiments like it, we have worked out pretty much the entire history of the universe. After more than four centuries of arguments based on speculations and prejudice, we now have data. We are living in the Golden Age of cosmology.

Because of that, we might be forgiven for looking around in awe at this assembly of the 'expert and the god'. These are, after all, the people who have given us an astonishing perspective on the universe, a perspective that humans have dreamed of since the time of the ancient Greeks. However, their story serves as a useful primer for what we are about to learn about science. Don't be fooled into thinking that their discoveries are part of a smooth progression in our knowledge.

WMAP examines the details of microwave radiation known as the cosmic microwave background, or CMB. The first prediction that a Big Bang would fill the universe with this type of radiation was made in 1948, just after the end of the Second World War. And it was forgotten almost as soon as it was made.

At that time, most people didn't believe there was a start to the universe. To the majority of physicists, the universe simply existed, and always had. What's more, the new theory about microwave radiation was born out of a combination of particle physics and astronomy, and, although plenty of people knew about either particle physics *or* astronomy, almost nobody was well versed in both. As if that wasn't enough of a problem, looking for this radiation would require microwave knowhow – and that was still a specialist area. That's why it took two decades and a set of lucky breaks to work out the history of the universe.

In 1963, a couple of astronomers based at Bell Laboratories in New Jersey found the CMB radiation by accident. Arno Penzias and Robert Wilson had been given a microwave detector in the form of a 15-metre long, 6-metre wide horn antenna to investigate why distant galaxies were emitting radio waves. Their first task was to identify the amount of noise in the detector, to make sure that any signal could be properly identified. As it turned out, there was an annoying amount of noise – far more than they had expected. They tried everything to get rid of it, even going as far as shooting the pigeons that were nesting in the horn antenna and removing the accumulated droppings.

Eventually, while at a conference in Montreal, one of them mentioned the problem to another astronomer, Bernard Burke. Burke thought nothing of it until he happened to be sent a paper by some Princeton astrophysicists. The Princeton group were suggesting that if the Big Bang really had occurred, the universe ought to be filled with microwave radiation. It was Burke's job to decide whether the idea merited publication – whether it was novel, and whether the idea stood up. He failed on the first call: he didn't make the connection with the same prediction made nearly twenty years earlier. However, Burke did make a connection with the troublesome noise in Penzias and Wilson's microwave detector. He put the Princeton theorists and the Bell Labs researchers in touch with one another. The result of that collaboration became front page news in the *New York Times*, and earned Penzias and Wilson a Nobel Prize.

Brand Science presents itself as if it takes a series of cool and logical (but brilliant) steps, a graceful flow of ideas from concept to irrefutable proof. That is a long way from the truth. 'Nearly all scientific research leads nowhere – or, if it does lead somewhere, not in the direction it started off,' Peter Medawar once wrote in a typically off-message pronouncement.

Scientists have a habit of airbrushing science's greatest moments to smooth out the human wrinkles and flaws in the process of discovery. Ultimately, though, scientists did themselves a disservice when they dehumanised their field. No wonder we have had such trouble keeping schoolchildren interested in science.

Education is just the tip of the iceberg. It is also no wonder that governments ignore the advice of scientists with impunity: they have been reliably informed that scientists are meek and unlikely to kick up a fuss. And the scientists, keen to perpetuate the myth of the scientist as public servant, play right along. No wonder the media don't give these scientists much space or airtime: who wants to be presented with dry facts by people who are just not like the rest of us? No wonder science has never been a part of popular culture: for generations, people have been persuaded that science is not like anything else humans get up to. No wonder scientific progress is slow: most scientists have spent their entire careers convinced they shouldn't do anything dangerous or too different from whatever is going on in the laboratory next door. They also know full well that they would fail to get funding or ethical approval if they dared to break out of the straitjacket.

It is time to embrace the reality about science, and discard the fantasy – before it is too late. We are building a civilisation on the foundations of science, placing our faith in its ability to support our hopes and deliver our needs. So far, scientists have been lucky: their cover-up has not resulted in the disastrous lack of trust it could have engendered if exposed in a malicious sting. That luck won't last for ever, though. Perhaps Daniel Sarewitz has put it best. 'The leap of faith that spans the chasm between laboratory and reality must be replaced with a bridge,' he says, 'lest ... we look down and realise that there is nothing underneath our feet.'

The work of science is too precious, and – in this age of approaching environmental crisis – too urgent, to allow that to happen. But safe in the knowledge that the public can cope with truly human scientists, and empowered by the realisation that people no longer fear science, we can set scientists free to work in the way that gives them their best chance of making progress. As the first step towards this, we are now going to peek behind the curtain and take an honest look at the lengths to which scientists have to go in order to make a breakthrough. Be warned: like Stephen Hawking's tea-time routine, it's quite a spectacle.

I

HOW IT BEGINS

Dreams, drugs and visions from God

It was humankind's first trip away from home. On 21 December 1968 a Saturn V rocket blasted off, its crew headed for the Moon. While in lunar orbit, however, the view through the craft's window distracted the Apollo 8 crew from their scheduled tasks. 'Though Apollo crews were trained to observe and photograph lunar features,' recalled astronaut William Anders, 'our main "discovery" was the Earth.'

On Christmas Eve the astronauts saw the entirety of their planet for the first time. Grabbing cameras and jostling for position, they took three photographs, two in black and white and one in colour. These are the celebrated 'Earthrise' pictures, astoundingly beautiful and moving images of our home that have been credited with kick-starting the environmental movement.

Stewart Brand, then a young Californian radical, felt rather proud of this achievement. One chilly afternoon three years earlier, Brand had been sitting on a gravel-covered roof in San Francisco's North Beach district. He was high on 100 micrograms of lysergic diethylamide: LSD. The buildings beneath him curved with the Earth's surface, and Brand's mind wandered back to a statement he had heard a month or so before. The architect and inventor Buckminster Fuller was giving a lecture, and Brand listened with rapt attention to Fuller's extraordinary claim. The root of all human misbehaviour, he maintained, lay in the fact that people perceive the Earth as flat. If only we carried with us the knowledge that our planet is a round ball, isolated in space, an island in an inhospitable cosmos, perspectives would change, Fuller said. On that rooftop, a question began to form in Brand's mind. 'Why haven't we seen a photograph of the whole Earth yet?'

The next day, Brand printed the question onto hundreds of badges and posters and sent them to NASA officials, members of Congress, Soviet scientists, UN officials and anyone else with influence and a publicly available mailing address. Then he set up a stall at the Sather Gate, the famous entrance to the University of California at Berkeley, where he sold the badges for 25 cents each. 'It went perfectly,' Brand said. What he means by that is, he was noticed. The university authorities threw him off the campus, an event that was reported by the *San Francisco Chronicle* and launched him onto the TV news bulletin that evening.

Brand took the campaign on the road, performing 'street-clown seminars' on space and civilisation at all America's major universities. He made the authorities edgy – the country was immersed in the Vietnam War, and even peaceful protests and rallies were always in danger of boiling over. That was why NASA hired an investigator to find out whether Brand and his 'Whole Earth' campaign constituted a threat to the United States Government. Years later, the investigator made himself known to Brand. 'I checked you out,' he said. 'You seemed all right, so I wrote them that this was California, where people took strange notions.'

At the bottom of his report, the investigator added a postscript. It said, 'P.S. By the way – why *haven't* we seen a photograph of the whole Earth yet?'

Brand's campaign began in February 1966. By the end of 1967, he recalls, the photos began to appear. Eventually, the Apollo 8 Earthrise photos, with the lunar surface arcing through the foreground, fulfilled everything Brand had hoped to achieve.

The photo taken by William Anders has been called 'the most influential environmental photograph ever taken'. Whether Brand's acid trip really did kick off the environmental movement is moot. It is impossible to tell whether the Apollo 8 astronauts knew about Brand's campaign. Though it is hard to believe they didn't, they have never mentioned it in any of their publicised discussions of the Earthrise photographs; according to the astronauts, they took those photos as an impromptu reaction to the majestic sight of the Earth from the Moon. Brand was certainly never mentioned in any discussion, but we can assume that NASA would not look kindly on suggestions that the agency was even remotely influenced by the antics of a drug-addled hippy.

The whole episode raises an intriguing question. Can this be how science happens? The Earthrise photo spawned a movement that is now a scientific endeavour – perhaps the most important endeavour that science has ever taken on. Did it really begin in an impassioned and drug-induced moment of inspiration? If so, it would certainly provide a perfect example of how science moves forward.

Ask a scientist what the scientific method is, Medawar said, 'and he will adopt an expression that is at once solemn and shifty-eyed:

solemn because he feels he ought to declare an opinion; shifty-eyed because he is wondering how to conceal the fact that he has no opinion to declare.' Invariably, the scientist will say something like, 'Well, you have an idea, then you test it in an experiment.' It sounds so straightforward. But where does the idea come from? From everywhere and nowhere. From wherever. Anything goes. Science, it turns out, is alarmingly like California: it is a place where people take strange notions.

Albert Einstein is reported to have remarked on one occasion that 'the secret to creativity is knowing how to hide your sources'. Though there is no paper trail for the attribution, it makes sense. Fyodor Dostoevsky once wrote that nearly all clever people are afraid of being ridiculous, and Einstein, of all the great scientists, was perhaps the most in danger of being ridiculed for his sources of inspiration. As biographer Hans Ohanian has put it, 'he made his profound discoveries in the manner of a mystic'.

Einstein relied on inspirations that had no traceable source. Working everything out logically, by deduction, is 'far beyond the capacity of human thinking', he said. Looking back on his experience, and relating it to the history of science, he admitted that 'the great steps forward in scientific knowledge originated only to a small degree in this manner'. That is something that Kary Mullis, another drug-using Californian, would agree with wholeheartedly. Mullis won the 1993 Nobel Prize in Chemistry, and he says he couldn't have done it without LSD.

Late one Friday night in May 1983, Mullis was driving along a Californian highway. His girlfriend was asleep beside him in the passenger seat. As can happen when you are driving, his mind was not really on the road. 'DNA chains coiled and floated,' he says. 'Lurid blue and pink images of electric molecules injected themselves somewhere between the mountain road and my eyes.'

DNA has a glorious mystique these days, but really it is just a complicated molecule. You can think of it working rather like Velcro: it is composed of two strands that stick together. Unlike Velcro, DNA has four different kinds of hook, which are known as A, T, G and C – abbreviations of their chemical names. Each hook can link to only one other kind: A links to T, and G links to C. When they are strung along one strand of the molecule, the order of the hooks – A, C, C, G, T, A, and so on – dictates what the other strand must look like in order for it to stick. This order also encodes the instructions for making specific proteins, the building blocks of biology.

In a biological organism, the DNA must make copies of itself, and it does this by pulling the two strands apart and bringing in new hooks to pair with the exposed ones. In this way, each single strand generates a new partner.

Kary Mullis was a humble hook-maker: he was producing the chains of acids that constitute DNA's A, T, G and C molecules. Despite his relatively lowly position in the Cetus Corporation, a biotechnology company based near Berkeley, he would habitually apply his mind to the bigger picture. He liked to imagine that one day we would understand the four-letter alphabet of the genetic code well enough to find where the copying had gone wrong and mistakes had crept in. If you could read the copying mistakes that lead to diseases such as Huntington's disease or sickle-cell anaemia, for example, you could potentially correct them, or at least avert the problems they cause. That's why Mullis would spend time throwing around ideas about how that might be done.

It's not an easy task. A strand of human DNA contains around a billion of the A, C, G and T hooks, or 'bases'. That's a daunting amount of reading to be done, especially when the writing is of a size where the book fits inside the microscopic nucleus of a biological cell.

But, Mullis reasoned, you don't have to read it all at once. He could assemble a single synthetic strand of just twenty bases, pull apart the strands of the DNA under investigation, and see if his synthetic strand fitted anywhere along it. If it did, you could encourage that little synthetic strand to reproduce itself by giving it some more bases, and the right enzymes and conditions to do the job. Do it enough times and you'd have a beakerful of copies of your strand. In a process reminiscent of what happens when Alice takes a bite of the 'Eat me' cake in Wonderland, that fragment of DNA would grow from the microscopic scale to the proportions of our world. Then you produce another short synthetic strand with a different sequence of bases, and do it all again. Eventually, you would be able to read the entire genome.

Now, after the fact, it sounds a fairly simple idea; indeed, Mullis says he still doesn't understand why no one else had thought of creating the 'polymerase chain reaction', or PCR. But perhaps it's because no one else had been taking the right drugs. Mullis had taught himself to think in abstract, visual ways. He knew what he was trying to achieve, but he applied himself to the problem indirectly, allowing himself simply to float around, immersed in the liquid with the very molecule he was trying to read. How had he done this? Through his use of hallucinogenics.

Here is how he describes his Eureka moment:

> I was down there with the molecules when I discovered it: you know, I wasn't stoned on LSD but my mind by then had learned how to get down there. I could sit on a DNA molecule and watch the polymerase go by ... that's just the way I think. I can put myself in all kinds of spots and I've learned that, partially I would think ... through psychedelic drugs.

Mullis has always been open about his use of hallucinogens. He tried LSD for the first time in 1966, just a few months before it was made illegal in the United States. When the ban came into force, he and a few colleagues began to synthesise and use hallucinogens that were still legal. Mullis believes that drugs are an invaluable tool for opening the mind to otherwise inaccessible insights. In a BBC documentary, he makes it clear where his debt lies. 'What if I had not taken LSD ever: would I have still invented PCR?' he asks. 'I don't know. I doubt it. I seriously doubt it.' Taking LSD was 'a mind-opening experience ... much more important than any courses I ever took.'

He is far from alone in this. If you use an Apple computer or iPod, or play computer games, or have ever had to submit DNA for forensic or medical testing, you have benefited from the drug's unique properties. Steve Jobs, founder of Apple Computers, calls his experience with LSD 'one of the two or three most important things I have done in my life'. The Nobel Prize-winning biologist Francis Crick was 'fascinated by its effects' during his acid trips. And most of the pioneers of Silicon Valley were regular users.

In 1991, a reporter from the *San Francisco Examiner* visited Siggraph, the world's biggest meeting of computer graphics engineers. The convention is held every year in California and draws everybody in the business. During the event, the reporter asked 180 of these top-flight professionals two questions: Do you take psychedelics? If yes, is this important in your work? Every one of them answered yes to both.

The investigation had been prompted by an article that came out in that July's *GQ* magazine. It was entitled 'Valley of the Nerds', and described widespread drug use among the pioneers of computer graphics. The writer quoted Chip Krauskorp, then head of Intel's Human Interface Program. Intel was happy to employ drug-users, said Krauskorp, because they were 'very, very, very bright' and gifted workers. The fact that they used psychedelic drugs, or cannabis, was not an issue; Intel even helped them get through the company's drug-testing procedures.

The work of some Californian mathematicians was also drug-inspired. Ralph Abraham, now an emeritus professor of mathematics at UC Santa Cruz, described himself in the *GQ* article as 'a purveyor of psychedelics to the mathematical community'. He later explained that he was a professor at Princeton in 1967 when he first tried LSD. His positive experiences prompted the move to California and eventually led him to work on the mathematics of computer graphics, then chaos theory and fractal geometry. 'There is no doubt,' Abraham says, 'that the psychedelic revolution in the 1960s had a profound effect on the history of computers and computer graphics, and of mathematics.'

In April 2008, the evolutionary biologist Jonathan Eisen announced on his blog that scientists were about to be forced to undergo anti-doping blood tests. The US National Institutes of Health, he said, was seeking to curb the growing problem of 'brain-boosting' drugs that enabled scientists to think more clearly in the pursuit of scientific breakthroughs. Eisen quoted an NIH official as saying that the new initiatives were 'designed to level the playing field among scientists in terms of intellectual activities'. The use of drugs 'has been affecting the competitive balance in scientific research,' according to the statement posted on Eisen's website.

The post was eventually revealed as an April Fool's joke, but not before several scientists had contacted Eisen to express their concern about when the testing would start. Scientists, you see, do take drugs.

In the world of the arts, drug use is, if not encouraged, hardly a scandal. Artists, writers and musicians have long appreciated that certain drugs can open the mind to new sources of inspiration and creativity. Jacob Bronowski believed that science is creative too – perhaps even more so than the arts. 'If any ideas have a claim to be called creative, because they have liberated that creative impulse, it is the ideas of science,' he said. But to be creative, scientists need ideas. And they, like artists, will take inspiration wherever they can find it.

Eisen's practical joke was stimulated by papers published in the journal *Nature*, which revealed that drug-taking is rife among scientists. In an article entitled 'Professor's Little Helper', two Cambridge University researchers announced that 'we know that a number of our scientific colleagues in the United States and the United Kingdom already use modafinil to counteract the effects of jetlag, to enhance productivity or mental energy, or to deal with demanding and important intellectual challenges'. The revelation provoked *Nature* to carry out its own informal survey of drug use among its readers. There were 1,437 respondents, largely drawn from the scientific community. A full 20 per cent of them admitted to using brain-enhancing drugs such as Ritalin (methylphenidate) or Provigil (modafinil).

Commentators looking to play down the findings pointed out that the effects of these drugs are mild, and that they were largely being used to get scientists through writing laborious grant proposals or long meetings. Most weren't using the drugs to help with the process of actually doing science. If that's true, it's a shame; such anarchy would undoubtedly speed the process of discovery.

John Maynard Keynes once stated that what made Isaac Newton great was his ability to focus his mind on a problem, and hold that focus until he had thought his way through it. 'I fancy his preeminence is due to his muscles of intuition being the strongest and most enduring with which a man has ever been gifted,' he said. Imagine, then, what Newton could have achieved on methylphenidate or modafinil. Modafinil is a stimulant that can help stave off the need to sleep. Methylphenidate – better known by one of its brand names, Ritalin – is generally used as a treatment for attention deficit hyperactivity disorder (ADHD) and helps the brain stay focused. Scientists less gifted than Newton certainly stand to gain something from their use. But perhaps it is the example of American psychologist and philosopher William James that they should really follow. Intoxication can be invaluable in releasing what lies beneath the conscious mind.

James carried out many of his investigations under the influence of drugs – in particular, nitrous oxide, or laughing gas. Intoxication opened his mind, he said: 'Our normal waking consciousness, rational consciousness as we call it, is but one special type of consciousness, whilst all about it, parted from it by the filmiest of screens, there lie potential forms of consciousness entirely different.'

In 1921, the German pharmacologist Otto Loewi had a dream that brought the field of neuroscience into existence. It was more than a hundred years since Luigi Galvani had shown that electrical impulses could contract muscles in the legs of frogs. A few decades later, scientists discovered electricity within biological tissue, and tracked it down to the nerve tissue. By the end of the nineteenth century it was known that nerves carried electricity, and that there were gaps between nerve cells; it seemed obvious to most researchers, working as they were in the age of the electric telegraph, that the signals between the nerves were electrical in nature too. Loewi and a few other researchers had suggested chemical transmission – chemical hormones were known to act as messengers in the body – but the idea was largely dismissed. Then came Loewi's dream:

The night before Easter Sunday of that year I awoke, turned on the light, and jotted down a few notes on a tiny slip of paper. Then I fell asleep again. It occurred to me at 6 o'clock in the morning that during the night I had written down something most important, but I was unable to decipher the scrawl.

The day dragged by mercilessly. Loewi spent it trying to remember the dream. He had no success, and put himself to bed early. At three in the morning the idea returned:

It was the design of an experiment to determine whether or not the hypothesis of chemical transmission that I had uttered 17 years ago was correct. I got up immediately, went to the laboratory, and performed a single experiment on a frog's heart according to the nocturnal design.

The experiment Loewi got up and performed in his laboratory before dawn is now a classic. He isolated two frog hearts. Into the first one he put Ringer's solution, a liquid which absorbs salts from the body until their concentration in the solution matches that in the surrounding tissue. Then he stimulated the heart's vagus nerve. As expected, the heartbeat slowed. Loewi then transferred the Ringer's solution to the second heart, whose nerves had been removed. On receiving the liquid, and thus the first heart's chemical salts, the second heart also slowed.

Loewi repeated the process, this time stimulating the accelerator nerve to speed up the first heart. When he transferred the Ringer's solution, the second heart began to beat faster. The experiment was an unqualified success, and overturned – overnight – the idea that the signals that pass between nerves must be electrical in nature. Transmission between nerves was a matter of chemistry.

According to Henry Dale, with whom Loewi shared a Nobel Prize, the discovery 'opened a new vista' in biology. Dale took Loewi's work and widened its scope, showing that all communication between nerves is chemical. This is the foundation on which modern neuroscience is built.

Finding inspiration during sleep is not an uncommon experience for artists; it merely speaks of the role of the subconscious. Paul McCartney woke up with 'Yesterday', one of the Beatles' most haunting and enduring songs, in his head, though the dream did not provide the finished lyrics – at first, McCartney had to content himself with rhyming 'lovely legs' with 'scrambled eggs'.

For scientists, dreaming a discovery seems – from the outside at least – to be much more uncommon; Loewi's tale has been called 'one of the most remarkable narratives of scientific discovery'. But it is not unique. The chemist August Kekulé won a Nobel Prize and made another hugely significant breakthrough after two separate dreams. The first, in 1855, occurred while he was travelling on a London bus. 'The cry of the conductor: "Clapham Road", awakened me from my dreaming; but I spent a part of the night in putting on paper at least sketches of these dream forms,' Kekulé said. 'This was the origin of the "Structural Theory".'

All today's vast chemical establishments, from DuPont to the myriad pharmaceutical companies, owe a debt to this dream because it gave Kekulé the secret of molecular structure. It is a measure of the importance of this breakthrough that the names of many of Kekulé's contemporaries in Germany still have a resonance in modern ears. His young assistant Adolf von Baeyer went on to win the Nobel Prize in Chemistry in 1905 and to found the Bayer pharmaceutical company. In the years that followed Kekulé's insight, Emanuel Merck managed to turn his father's apothecary in Darmstadt, where Kekulé grew up, into the multinational pharmaceutical company that now supplies drugs to the world. Kekulé moved to Heidelberg to be close to Robert Bunsen, the man who invented the Bunsen burner still in use in every chemistry laboratory in the world. Thanks to Kekulé's dream, this group of chemists laid a foundation that is still strong 150 years on.

Not that Kekulé said anything about his dreams to these colleagues – he revealed his sources only a few years before his death. This seems to be a recurring theme. Einstein, too, kept the strange inspiration for special relativity locked away until he had finished in science. Only in his autobiography, written in his final decade, did Einstein reveal that, at the tender age of sixteen, he had experienced a vision. He saw himself running alongside a beam of light, an experience that conjured up a puzzle in the young Einstein's mind. He visualised the light as an electromagnetic wave composed of two oscillating fields – one electric and one magnetic – just as most physicists of the time would have done. Normally, such a wave would rush past him at great speed – the speed of light – but in the vision it merely stretched out in front of him.

Two things struck him immediately. First, if you run alongside a light beam, you see the waves as stationary. Such a stationary electromagnetic field would not correspond to anything we would experience as light. Second, Einstein instinctively knew that he ought to experience nothing different in his interactions with the world around him, just because of motion.

The situation is no different to finding yourself inside a train carriage, travelling through a pitch-black night at an unwavering speed. There can be no way for you to know that you are even moving – there is no experiment you can carry out that will reveal your motion relative to the landscape outside the window. Moving with the light beam, with nothing else around, Einstein saw that there would be no way for him to tell that he was travelling at the speed of light – or at any other speed, for that matter: 'every-thing would have to happen according to the same laws as for an observer who, relative to the earth, was at rest', he said.

But the implications of his vision said otherwise. According to the standard electromagnetic theory, the laws of physics *would* depend on your speed: if you travelled fast enough, you *would* experience something as nonsensical and resistant to analysis as a stationary light wave. Einstein wrote in his autobiography that 'the germ of special relativity theory was already present in that para-dox'. When he constructed the theory a decade later, he resolved the problem entirely by declaring the speed of light to be constant, whatever the speed of the source of the light. All the strangeness of relativity – the elastic nature of time and distance, for example – follow from this insight.

In revealing his sources only at the last minute, Einstein was following a grand tradition. Is the fact that the sources of ideas are so 'unscientific', so irrational, somehow embarrassing to scien-tists? The truth is, for most scientists there is nothing to be gained from revealing the inspiration. And there is much to lose: the joy of bamboozling your colleagues, for instance. The Renaissance sci-entist Girolamo Cardano, for one, took great pleasure in deceiving his colleagues into thinking he was much cleverer than they.

Cardano was born in Pavia, Lombardy in 1501, the illegiti-mate child of a lawyer friend of Leonardo Da Vinci. He arguably contributed even more to science than Da Vinci. The driveshaft of your car employs a pivoting joint that allows rotary power to be transmitted at an angle. This joint is known as a Cardan joint in honour of its inventor. Cardano also published more than a hundred books that ranged across mathematics, natural sciences, medicine, engineering and philosophy. It was he who came up with the mechanical gimbal that made high-speed printing possi-ble. He also originated the study of probability – largely because of his interest in gambling. Perhaps most impressively, he developed the concept of imaginary numbers, a vital part of the theories of electromagnetism, quantum physics and relativity that were still hundreds of years away.

The year before he died, Cardano wrote an autobiography entitled *The Book of My Life*. It is an electrifying read and details his sexual conquests, his illnesses, his tragedies and – perhaps most enlightening of all – his *modus operandi* for scientific discovery. Cardano seems to have taken great delight in the assumption that his breakthroughs came from rational sources. 'This form of knowledge is pleasing to the erudite, for they think it proceeds from great learning and practice, and on this account very many have judged me to be deeply devoted to study and possessed of a good memory when nothing is less true,' he wrote.

The truth – in Cardano's eyes, at least – is that his source was 'the ministrations of my attendant spirit'. He was visited by familiar spirits, angels and demons, and took advice from them: 'I have less rarely arrived at comprehension by a skilful treatment than I have been aided on many occasions by spiritual insight'. Cardano's father, son and cousin were all certified lunatics. Cardano called himself a lunatic. Perhaps we too would question his sanity. But the fact is that invisible, intangible, irrational sources can be an incredibly powerful source of discovery.

The strange story of the genesis of the electric motor, as hallucinated by the twenty-two-year-old Nikola Tesla, provides further proof. One afternoon in 1881, Tesla was out walking with some student friends in Budapest Park. They were walking west, towards the setting sun. Tesla, entranced by the spectacle, was reciting a poem by Goethe:

> The glow retreats, done is our day of toil;
> It yonder hastes, new fields of life exploring;

> Ah, that no wing can lift me from this soil,
> Upon its track to follow, follow soaring …

He stopped mid-stanza. In front of him was a vision of a fiery magnetic field being made to rotate by a ring of electromagnets. The magnets were powered by a current that varied in the pattern of a sine wave, with a delay between the phase of current supplied to each magnet, so that the field moved in a circle – just as a ring of lights turned on one after the other give the appearance of a light rushing around the loop. Inside the ring of magnets, Tesla saw a hulk of iron that could be connected to the electromagnets in one way to start it spinning in one direction, and in another way to reverse its motion. He had a moment of catatonic immobility, then blurted out, 'See my motor here … watch me reverse it.' His friends grabbed him and shook him until he came back into their world.

When Tesla returned to the laboratory and built what he had seen – what is now known as the self-starting alternating current motor – it worked first time.

The resolution of the Second World War also owes a debt to an unbidden hallucination. To make an atomic bomb requires weapons-grade plutonium. This can be produced in a nuclear reactor, but only when you have worked out how to protect the reactor's uranium from the corrosive effects of the water it heats. Somehow, the uranium has to be encased in a shell thick enough to make it submersible, waterproof and gas-tight, but not so thick that it absorbs all the heat. Nor can it absorb the neutrons emitted by the uranium, because that would kill the chain reaction so essential to the process.

According to the official report of the Manhattan Project, the US effort to build an atomic bomb, this 'canning' of the uranium was one of the most difficult problems they faced. It had been holding up the development of the bomb for months. There were just a few weeks remaining before uranium was scheduled to be loaded into the reactor, and tempers at the Project were becoming more frayed by the day. This was 1943, and everyone knew that Hitler's scientists were also chasing the bomb. The outcome of the war was likely to depend on who got there first.

Then one day, 'one pace past the water cooler', as he recalled, physicist Omar Snyder saw the can and how to make it. 'The entire process for the manufacture ... flashed in my mind instantaneously,' Snyder said later. 'I didn't need any drawings; the whole plan was perfectly clear in my head.' Immediately he went to his lab and began work on turning this revelation into a reality. Within a day and a half it was finished, and Snyder had single-handedly – and inexplicably – solved the problem that was holding up the production of the US atomic bomb.

Would other scientists have got there eventually, without such an irrational inspiration? It's not clear that they would. The British had tried and failed for years, as had the Russians. The successful Russian reactor design is a carbon copy of the American prototype, obtained through espionage rather than inspiration.

Snyder said that he didn't believe there was anything about this incident that made him special: similar processes of revelation were happening all around him. In fact, several major developments in physics that made the bomb possible occurred as a result of an irrational, unpredictable – some would say unscientific – moment of revelation or inspiration.

Take the experience of Snyder's boss at the Manhattan Project, for example. The physicist Enrico Fermi, a defector from Mussolini's Fascist Italy, was the man charged with building the nuclear reactor. Back in Rome, in October 1934, Fermi had experienced a similarly inexplicable breakthrough – one that led directly to the atomic bomb. He had been trying to understand nuclear reactions. He and his team were trying to induce radioactivity by bombarding a metal target with neutrons, but there was neither rhyme nor reason to their results: there seemed to be no way to predict what would happen in any particular experiment.

One day, on the way into work, Fermi decided to try putting a block of lead in front of the target. The notion in his head was that the lead might filter out the slowest of the neutrons and allow a more controllable bombardment. He commissioned the physics department's machinists to make the lead block. Contrary to his usual style, he gave them an extremely detailed description of what he wanted, and the block was fabricated to his requirements. What happened next was, as with Snyder's experience, inexplicable – and yet of phenomenal consequence.

For some reason he couldn't fathom, Fermi felt that he didn't want to use the lead block. For two days he let it sit, awaiting collection, in the machine shop. When he eventually had it brought to his laboratory, he still dithered. This is how he later recounted the story to the astrophysicist Subrahmanyan Chandrasekhar:

> I was clearly dissatisfied with something: I tried every 'excuse' to postpone putting the piece of lead in its place. When finally, with some reluctance, I was going to put it in its place, I said to myself: 'No! I do not want this piece of lead here; what I want is a piece of paraffin.' It was just like that: with no advanced warning, no conscious, prior, reasoning. I immediately took some odd piece of paraffin I could put my hands on and placed it where the piece of lead was to have been.

Using the paraffin turned out to be an extraordinary stroke of anarchic genius. Fermi and his collaborators watched as the radioactivity induced in their silver target rose by 50 per cent. They soon realised that the hydrogen molecules in the paraffin were slowing the neutrons and giving them far more opportunity to interact with the silver atoms – creating radioactive elements – before they emerged on the other side. Such slow neutrons are now known to be the essential component of a reliable nuclear reaction. Fermi's discovery won him the 1938 Nobel Prize in Physics and, arguably, it won the Allies the Second World War.

Were Snyder and Fermi's experiences unusual? Yes, in that scientists do not make such significant discoveries every day. But if we restrict ourselves to the realm of significant scientific discoveries, the answer seems to be no: they invariably come from apparently nowhere.

Many of the scientists who have experienced such revelations feel humbled by them, and even question whether they should receive credit for the discovery; to them, it seems more like a dispensation of grace from on high. The physiologist Alan Lloyd Hodgkin once referred to a 'feeling of guilt about suppressing the part chance and good fortune played' in his Nobel Prize-winning work. The British mathematician Paul Dirac was afflicted by something close to guilt over the ownership of some of his ideas, which he described as coming 'out of the blue'. 'I could not very well say just how it had occurred to me,' he wrote in 1977. 'I felt that work of this kind was a rather "undeserved success".' Michael Faraday apparently felt the same. He repeatedly turned down a knighthood for his achievements, preferring to remain 'plain Mr Faraday'. The source of Faraday's humility was that all his inspiration and discovery came from his faith in God.

'I am of a very small and despised sect of Christians, known, if known at all, as Sandemanians.' That was how Michael Faraday introduced himself to Ada, Countess of Lovelace, in 1844. The Sandemanian sect was a group defined by a particularly strict adherence to the precepts of Christianity as laid down in the New Testament – and a particular abhorrence of any other Christian group. Belonging to any national or historical body, such as the Church of Scotland or the Roman Catholic Church, was seen by the Sandemanians as gross error.

The Sandemanians enforced a strict code of conduct, and enthusiastically followed the New Testament recommendation of throwing out of their church anyone discovered to be engaging in sinful acts – including, as one set of church records notes, such transgressions as 'not being sufficiently humble'. There was no such thing as a half-hearted Sandemanian, and the fact that Faraday spent two periods of his life as a Sandemanian church elder indicates the fervour of his faith.

But, surprising as it may seem to modern eyes, it was not a faith that saw science as something to be treated with suspicion. To the Sandemanians, the New Testament gave a clear mandate for science. In his Epistle to the Romans, St Paul observes that, 'since the creation of the world God's invisible qualities – his eternal power and divine nature – have been clearly seen, being understood from what has been made, so that men are without excuse'.

Faraday quoted this passage twice during his public lectures. It was his calling, as he saw it, to study nature, which was 'written by the finger of God', and make clear the eternal power and divine nature of the Creator. That, after all, was how the people could turn to Him and be saved. As Faraday put it, 'unravelling the mysteries of nature was to discover the manifestations of God'. Small wonder, then, that he seemed so unmoved by the technological applications of his discoveries: his calling was to expose the laws of nature, and thus the nature of God. What others did with his discoveries was of no concern to him.

The son of a blacksmith, Faraday's involvement with science came about when he worked as a bookbinder's apprentice. He was intrigued by the contents of the scientific books he was charged with binding, and read them voraciously. He landed a job at the Royal Institution at the age of twenty-one, after approaching Humphry Davy, being turned down, then calling back after Davy's assistant was sacked for fighting in the main lecture theatre. It was a remarkable stroke of good fortune for the Royal Institution: Faraday proved himself a meticulous and brilliant experimental scientist.

At the time, Europe's physicists were exercised by the nature of electricity. It was known that similar electric charges repelled each other, and that an electric current produced a magnetic field. With these discoveries fresh off the scientific presses, Faraday's friend Richard Phillips asked him to produce a historical account of the breakthroughs for the journal *Annals of Philosophy*.

Faraday, with a diligence that would become his trademark, was not content simply to read and digest every paper published on the subject of electricity: he also recreated every experiment. By the time he was ready to write his account, he had an intimate knowledge of the theoretical and experimental limits of this branch of science. He realised that finding the link between magnetism and electricity would be the key to making progress in this area. And he had his own unscientific prejudices that would help him do just that. If God said He had made Himself known through nature, then nature's laws must be comprehensible. 'I believe that the invisible things of HIM from the creation of the world are clearly seen,' Faraday declared. All he needed to do was to find out where to look.

Perhaps that makes it sound trivial. It was not. Science is not straightforward or obvious; it is not about simply collecting enough evidence to prove a point. It is about making connections. Many studies have shown that you can hand scientists all the evidence they need to make a breakthrough, but there is no guarantee that the breakthrough will come.

André-Marie Ampère, the French scientist whose contribution has been honoured in the name given to the unit of electrical current, came at the mysterious connection between magnetism and electricity with a furious mathematical bent. But Ampère was convinced that electricity was the flow of some kind of fluid within wires, and that this flow could be modelled mathematically to expose the origin of magnetism. It got him nowhere. Faraday, on the other hand, had a simpler angle of attack: the nature of God.

To this blacksmith's son, mathematics was a foreign language anyway. Faraday quickly got lost when he tried to follow Ampère's arguments. ' With regard to your theory,' Faraday wrote to Ampère, 'it so soon becomes mathematical that it quickly gets beyond my reach.' And so plain Mr Faraday had to find another way forward.

The Sandemanian reading of the Bible left Faraday with a series of impressions and intuitions about how the physical world would be influenced by its Creator. In 1844, for example, he wrote a note on the nature of matter, speculating about atoms: 'by his word' God could have spoken 'power into existence' round points in space, he said.

Faraday was almost entirely alone in the scientific world in considering the role of empty space. For trained scientists, the laws of attraction and repulsion between electrical charges held at a distance, and the equivalent point-to-point gravitational attraction between masses, were as natural as breathing. As a result, they thought only in terms of the influence of something at one point on something else some distance away.

But the Bible clearly stated that God filled all of space. Isaiah's vision of the angels who call out 'Holy, holy holy is the Lord God Almighty; the whole earth is full of his glory' is just one of the passages that the Sandemanians claimed as a denial of any emptiness in the universe. To Faraday, intent on discovering the nature of God through His influence on the physical world, this seemingly empty space had to be of interest. It also made sense to Faraday that such divine influence *could* be discerned. The properties of matter 'depend upon the power with which the Creator has gifted such matter', he said.

According to science historian Geoffrey Cantor, Faraday saw himself as investigating a 'perfectly designed system' in which all events are tightly ordered by divine providence and held in a self-sustaining system with matter and force both conserved. Forces can be transformed into one another, but neither created nor destroyed by any human power.

Added to this was Faraday's concept of symmetry: cause and effect, action and reaction, north and south. For him, everything in nature was somehow correlated with something else. And it was all subject to the law of 'unity in diversity'. In his First Epistle to the Corinthians, St Paul said that 'There are different kinds of service, but the same Lord.' The lesson, to Faraday, was clear. 'Like the members of the Sandemanian community who work in harmony for the common spiritual good,' Cantor wrote, 'so the different material bodies and the laws of nature cooperate with one another within the system of nature.'

All of this led Faraday to a particular view – a preconception – of what he would discover in his experiments. First he discovered the magnetic field that inhabited 'empty' space around a magnet. His view of the integrity of all things led him to conceive this field as being composed of closed loops: for him, circular forms were more reflective of the Creator than were lines that stretched merely from one point to another.

The discovery of electromagnetic induction came from a similarly spiritual part of Faraday's mind. This is the phenomenon whereby the movement of metal wire within a magnetic field generates electricity in the wire. This was revolutionary to the trained mathematical scientist, but it all made perfect sense to Faraday. The phenomena of electricity and magnetism were tied together in mutual embrace. If a moving electrical conductor produced a magnetic field, then a moving field would be expected to produce a current in a conductor. Movement, magnetism and electricity were a reflection of the Trinity: locked together, separate but inseparable – a mystery.

It is thanks to this faith-inspired discovery that you have electrical power delivered to your home. And thanks to the symmetry of nature, Faraday showed that we can turn the arrangement on its head, allowing a current to flow within a magnetic field to create motion. Here we have the genesis of yet another staple of the modern world: the electric motor that powers everything from giant industrial plant to computer disk drives.

Faraday was by no means the only scientist to be motivated by a religious faith. Nicholas Copernicus, who removed the Earth from the centre of the universe, referred to nature as 'God's Temple' and claimed that God can be known through the study of nature. It is ironic that the same attitude, which would be heavily criticised if it were raised in scientific circles today, also caused Copernicus's work to be put on the Catholic Church's *Index Librorum Prohibitorum*, its list of forbidden books.

Inspired by Copernicus's orbiting planets, and adopting the same view as Faraday – that God would use a system of 'unity in diversity' – the surgeon William Harvey theorised that the human body had a circulatory system that mirrored the orbits of the planets. 'I began to think whether there might not be a Motion, As It Were, In A Circle', he wrote in 1628, when he revealed the results of his investigations into the movement of blood around the body.

His conclusion was that the heart 'is the beginning of life; the sun of the microcosm, even as the sun in his turn might well be designated the heart of the world'.

God is not popular in science these days. A survey of members of the US National Academy of Sciences revealed that 85 per cent reject the notion of a 'personal God'. Even that is not good enough for some scientists. The astronomer Neil de Grasse Tyson, for example, has turned the statistic on its head, lamenting that 15 per cent of 'the most brilliant minds this nation has' accept the idea of a personal God. 'How come that number isn't zero?' he asks.

Religion is also a thorn in the flesh of Oxford University chemist Peter Atkins. His response to the NAS survey was similar to Tyson's. 'You clearly can be a scientist and have religious beliefs,' he told the *Daily Telegraph*. 'But I don't think you can be a real scientist in the deepest sense of the word.' For Atkins, religious belief and a scientific worldview are mutually 'alien categories of knowledge'.

However, the evidence doesn't bear Atkins' statement out. Michael Faraday, for one, stands as proof that holding religious beliefs – as well as taking drugs and experiencing dreams, visions and moments of 'undeserved' insight – can provide the key to scientific discovery. Why? Because science is much more irrational than scientists would like to admit.

If there is a common thread to be teased out of these tales of drugs, dreams, visitations and visions, it is surely that, as Feyerabend said, 'anything goes' for the scientist. It certainly seems that 'anything comes': inspiration appears unbidden from the most unlikely of sources. James and Mullis had their drug-induced experiences, Einstein and Tesla their exuberant visions, Loewi and Kekulé their dreams, Faraday his faith ... There is no sense in ruling out any approach to a scientific question.

Kekulé suggested to his colleagues that, in search of breakthroughs, they should 'learn to dream!' We might equally encourage scientific progress by learning to take drugs, or embracing a worldview where a Divine Hand has created a rational, intelligible universe. After all, as that old mystic Einstein said, 'The eternal mystery of the world is its comprehensibility.' If there is nothing rational, logical or obvious about the universe's openness to investigation, why would our methods of investigation need to be any different?

There is a secret anarchy, then, behind the inspirations of science, but it is nothing compared with the anarchy that follows. After the initial inspiration, the scientist has to gather evidence to confirm or refute the idea. This process is the bedrock of science, and it is why scientists throw up their hands in horror or shame whenever a claim of scientific fraud surfaces. But this seems to be something of a pantomime. It must be – why else would scientists refer to fraud as 'normal misbehaviour'?

2

THE DELINQUENTS

Rules are there to be broken

On a clear, cold morning in January 2008, a group of students walked nervously through the campus of La Sapienza, Rome's oldest university. When they reached the university's centrepiece, a huge bronze statue of Minerva, they took a nervous look over their shoulders and set to work. They taped a banner to the pedestal beneath Minerva's skirts and stepped back to admire their moment of anarchy. 'Knowledge needs neither fathers nor priests,' the banner declared. 'Knowledge is secular.'

The message was a direct challenge to the Vatican. Later that week, Pope Benedict XVI was due to cross Rome to visit La Sapienza, and the students and faculty were far from happy about it. The Pope, they said, was 'anti-science'. Elsewhere on the campus, others were making their feelings known in different ways.

Student protestors had taken over the Rector's office. Scores of faculty members had signed a letter, published in the daily newspaper *la Repubblica*, voicing strong objections to the visit. The Pope's presence at the university would be 'incongruous', the letter said.

The anarchists won the battle: that evening, Cardinal Tarcisio Bertone, the Vatican's Secretary of State, sent word that the visit was to be cancelled. Fearing a humiliating scene, Bertone regretted that 'the conditions for a dignified and peaceful welcome were lacking', and conveyed his apologies to the Rector. The news was greeted with a resounding cheer from students and professors. Then, a few hours later, their delight melted into awkward embarrassment. It turned out they had got the Pope all wrong.

The demonstrations had been prompted by a speech delivered by the Pope back in 1990, when he was Cardinal Ratzinger. The basis for the letter in *la Repubblica* was the transcript of this speech, taken from the Italian-language Wikipedia website. The Cardinal, the letter said, had defended the Church's decision to put Galileo on trial for suggesting that the Earth moves around the Sun. The writers of the letter damned the Pope with his own words: he had described the trial of Galileo as 'rational and just', they said, and declared that 'the Church at the time of Galileo was much more faithful to reason than Galileo himself'. The sixty-seven signatories made their feelings very clear: 'These words offend and humiliate us.'

We might reasonably expect those signatories, as scientists, to have checked their facts. If they had called up the Wikipedia page for themselves, and scrolled down to Ratzinger's discussion of the Galileo affair, they would have seen that the Cardinal was not attacking science – quite the opposite, in fact. He was attacking those who stood by the medieval Church's attack on Galileo.

Ratzinger singled out one man for particular criticism: the twentieth-century philosopher Paul Feyerabend.

In his 1975 book *Against Method*, Feyerabend examined the case of Galileo vs Pope Urban VIII and came to what is, in the modern age, a rather surprising conclusion. Given the nature of the scientific evidence, the robustness of the argument, and the ethical and cultural implications of Galileo's claims, the arrest and conviction of Galileo was 'rational and just', Feyerabend says. 'The Church at the time of Galileo was much more faithful to reason than Galileo himself.'

The quotes the La Sapienza professors had eagerly attributed to Ratzinger were actually from Feyerabend. And, as anyone reading the whole of Ratzinger's 1990 speech could see, Ratzinger, after quoting *Against Method*, declared Feyerabend's conclusions 'drastic' given that the philosopher knew full well that Galileo had been right. What's more, responding to hardliners who suggested that the Church should have been even tougher on Galileo, Ratzinger went on to declare that the faith 'does not grow from resentment and the rejection of rationality'. According to Giorgio Israel, a Jewish mathematician who wrote a commentary on the drama in the Vatican's own newspaper, the speech 'could well be considered, by anyone who read it with a minimum of attention, as a defence of Galilean rationality'.

The La Sapienza professors had based their opposition to the Pope's visit on unverified and fallacious claims that served only to confirm their prejudices. As this embarrassing truth emerged, several of the sixty-seven signatories – Luciano Maiani, the physicist who heads Italy's main scientific research body, for instance – sheepishly withdrew their objections to the Pope's visit.

Feyerabend, who died in 1994, would have been amused and delighted by the goings-on at La Sapienza. The manufactured outrage of the university's scientists is a perfect illustration of his most treasured idea: that scientists are anarchists who pay no attention to rules and 'accepted practice'. After all, the La Sapienza professors were not the first scientists to justify their prejudices with unquestioning use of convenient evidence. Einstein played the same game. Other Nobel laureates, such as Robert Millikan, have done it. Ptolemy, Newton and the beloved Galileo have also been found guilty of progressing science by taking a flexible approach to their experimental observations. Today's scientists are no different. In 2006 the journal *Nature Cell Biology* declared in one of its editorials that as many as one in five of its accepted papers contained 'questionable data' – even after the journal had introduced a data-screening process.

But to scientists, data are not really to be trusted anyway. When Francis Crick and James Watson were hunting down the structure of DNA, they had to sweep aside the 'truths' that others had found. Their crucial breakthrough came when a colleague looked over their shoulders and remarked that the textbooks they were slavishly following contained information that was just plain wrong. They were being misled by guesses (about the angles of chemical bonds) that had been repeated so often that they had gained the status of fact. As a result, Crick said, he learned 'not to place too much reliance on any single piece of experimental evidence'. Watson's view was similar: 'some data was bound to be misleading if not plain wrong'. Crick and Watson could not have made their world-changing discovery without developing this attitude. When it comes to data, scientists have to be anarchists. And it has ever been thus.

Historians of science attribute the earliest scientific fraud to the Egyptian mathematician and astronomer Ptolemy: in the second century AD, he manipulated data to support his astronomical models. Some scientists, though, didn't have the luxury of manipulating the important data. Galileo, for example, just had to hope that sheer force of personality would be enough to stop people noticing his sleight of hand.

Perhaps Galileo's inclination towards anarchy should have been obvious all along. Though deeply religious, he brought three children into the world *in fornicazione*, as the parish register of San Lorenzo puts it. The mother of all three – two daughters and a son – was Galileo's lover Marina Gamba; for reasons that no one has ever understood, Galileo never married Marina. This rebellious relationship seems to foreshadow his more direct, and more famous, challenge to the traditions of the Catholic Church.

The free-thinking Galileo must have been pleased when Maffeo Barberini became Pope Urban VIII in 1623. Caravaggio's portrait of the man has a playful air, and the new pontiff was reportedly something of a Renaissance man. He had been a supporter of Galileo's scientific efforts, and enjoyed discussing Galileo's ideas. One of the matters under discussion was Copernicus's heliocentric model of the universe, in which the Sun, not the Earth, was the hub of the cosmos. Galileo was keen to prove Copernicus right, and Urban VIII was keen to hear a convincing argument. The tides held the proof, Galileo said, and he proposed to the Pope that he write a book entitled *Dialogue on the Tides*. Urban saw the bigger picture, and insisted the book be called *Dialogue Concerning the Two Chief Systems of the World*.

The theory of the tides comes in the fourth chapter of the *Dialogue*, and provides what Galileo considered to be his most forceful and conclusive proof of the motion of the Earth through the heavens. His argument centres on two facts: if the Earth moves as Copernicus had suggested, then it has both rotational motion (through its spin on its axis) and linear motion (along a path through space). Any point on the Earth's surface therefore moves rather like a point on the rim of a cartwheel: around and along. This combination of movements produces an ever-changing speed of motion. And an ever-changing speed, as anyone who has carried a cup of ale in a horse-drawn cart knows too well, causes sloshing. Here, Galileo said, is the cause of the back-and-forth motion of the tides.

Except that it isn't. The mathematics of Galileo's theory creates one tide per day, and as any of his Venetian friends could have told him, there are two. Also, high tide should happen at the same time every day according to Galileo's calculations. It doesn't, as every sailor knows. Perhaps most heinously, Galileo makes no attempt to account for the Moon's involvement with the tides, even though its influence was well known at the time. Johannes Kepler had made this point three decades earlier, in a 1609 treatise called *Astronomia Nova*. Galileo, unwilling to let the Moon destroy his precious idea, resorted to mocking Kepler's openness to 'puerilities' about the Moon's 'occult properties'. The idea that the Moon had any influence on the tides was a 'useless fiction', he said.

It is stretching anyone's credulity to suggest that Galileo was not aware of the mismatch between his theory and what was then common knowledge about the tides. It seems he simply ignored inconvenient data. He was convinced – rightly – that the Earth moved, and he was prepared to try to convince others, by any means necessary.

Isaac Newton tried something similar. He is arguably the greatest genius that ever lived, and had 'a strength of mind almost divine', according to his monument in Westminster Abbey. He was the first man of science to be granted a state funeral, and his eminence and scientific brilliance were bright enough to cause Alexander Pope to pen these famous words:

Nature and Nature's laws lay hid in Night.
God said, 'Let Newton be!' and all was Light.

Pope says nothing, however, about Newton's dark side. He was a man with few friends and many bitter enemies – especially among those who dared to dispute his scientific claims. Dissenters were met with a barrage of insults and ferocious attacks on their character and work. Later in life, Newton took on the role of Master of the Royal Mint, and showed further disregard for the notion of truth. He was singularly vindictive in his bid to curb forgeries. Counterfeiting was then a treasonable offence, and the penalty for treason was to be hung, drawn and quartered. Newton condemned many to death on the flimsiest of evidence, sometimes nothing stronger than the word of a paid informant.

It has been suggested that the unpleasant character traits that became so obvious in Newton's later life can be blamed on his gradual ingestion of mercury during alchemical experiments. But it is clear that the dark side was there all along. Parts of Newton's most celebrated work – the *Principia* – are 'nothing short of deliberate fraud', according to his biographer Richard Westfall: 'If the *Principia* established the quantitative pattern of modern science, it equally suggested a less sublime truth – that no one can manipulate the fudge factor quite so effectively as the master mathematician himself.'

Newton fudged theoretical calculations of the speed of sound, the precession of the equinoxes, the strength of gravity on the Moon and the heights of the tides so as to fit with experiment. And in each new edition of the *Principia* he introduced changes that took the same data but significantly increased the level of apparent precision. Westfall calls this 'a cloud of exquisitely powdered fudge factor blown in the eyes of his scientific opponents'.

The thing is, to scientists it seems that this is all just fine. Ptolemy has been forgiven as 'honestly motivated'; there's nothing unusual about publishing only the data that support your theories, according to Harvard historian Owen Gingerich. No lesser a figure than Einstein has exonerated Galileo – this time because Galileo was right about the motion of the Earth around the Sun. 'It was Galileo's longing for a mechanical proof of the motion of the Earth which misled him into formulating a wrong theory of the tides,' wrote Einstein in a preface to a modern edition of the *Dialogue*. 'His endeavors are not so much directed at "factual knowledge" as at "comprehension".' And here is where we start to see a new pattern emerge, one that exposes the secret anarchists.

A 2007 report into scientific misbehaviour published in *Nature* concluded that 'many of the risk factors for misconduct also seem to be what makes for good science'. That certainly seems to be the case. Galileo and Newton were the founding fathers of science. Newton in particular made great play of the role of observations and data, setting the tone of science for centuries to come. But data, as we have seen, are not always reliable, and in private scientists rely on intuition to guide them in their work. When intuition and data clash, it is usually intuition that wins out. As Peter Medawar pointed out, 'scientists who fall deeply in love with their hypothesis are proportionately unwilling to take no as an experimental answer'.

Is this justifiable? Yes, if the object of their infatuation turns out to be worth the attention.

As the twentieth century began, Robert Millikan was fast approaching forty. All around him, physics was at its most exhilarating: J.J. Thomson had just discovered the electron, and Max Planck had pulled quantum theory into existence with a brilliant piece of scientific detective work. Outshining everyone else, Einstein had made it clear that everything was composed of atoms and, with his special theory of relativity, that the universe was stranger than anyone had imagined.

Millikan, on the other hand, had done practically nothing. So he decided to measure e, the charge on the electron. Finding the value of e was important because it – and the very existence of the electron – was the subject of a heated and complex international debate. Although Thomson had ostensibly discovered the electron in 1897, German physicists – at that time considered the best in the world – were unconvinced.

Their hesitations were to do with the aether, a ghostly fluid that was thought to fill all of space. The aether provided a medium through which light could travel, and in the corridors of Germany's universities it was agreed that the experiments which claimed to demonstrate Thomson's 'negatively charged matter' merely provided evidence that electricity was a strain in the aether.

According to modern accounts of the history of science, it shouldn't have been possible to hold this view in the twentieth century. In 1887, two American physicists, Abraham Michelson and Edward Morley, had performed an experiment which showed that the aether did not exist. They had been trying to measure how fast the Earth's motion through space was moving us through the aether by finding the direction in which light moved fastest. The direction of motion of a point on the Earth's surface is constantly changing as the Earth rotates and moves around the Sun. In the same way that you feel a wind when you move through the air, there should be an 'aether wind' as the Earth moves through the aether. And because the aether carries light, the Earth's motion should produce a measurable difference in the speed of light coming from different directions. To their surprise, Michelson and Morley found no such difference: light had no 'preferred' direction. The only explanation was that there is no aether.

Michelson and Morley's work is now cited as a classic experiment, but it has been given something of a positive spin. Because there is no aether, the plan to measure the speed of the Earth's movement through it was doomed to failure. In science, a null result is not always reported, let alone shouted from the rooftops, and this experiment was very easy to ignore. It took decades for this null result to come to the attention of the international community. No one is sure whether, ten years on, the German physicists had even heard of Michelson and Morley's experiment.

Millikan, though, would have been well aware of it: Abraham Michelson was Millikan's boss. It doesn't take great insight into human nature to surmise that Millikan knew his career would gain an immediate (and desperately needed) boost if he could measure *e*. If he could find the charge on an individual electron, he would cast more shadow on the aether theory, validating Michelson as well as Thomson. To a struggling junior researcher nearing the middle years of his career, the prospect must have been irresistible.

Millikan's idea was simple. A droplet of water that had been given an electric charge would be attracted to a metal plate which carried an opposite charge. He arranged his apparatus so that the electrical attraction pulled the droplet up, while gravity pulled it down. This gave him a way to measure *e*. First he would find the mass of the droplet by measuring its size. Then he would measure how much voltage is needed for the attraction to the metal plate to cancel out the downward pull of gravity. From those two pieces of information he could get a measure of the charge on the droplet. Millikan guessed – correctly – that whatever the charge on the droplet, it would always be some integer (whole-number) multiple of the same number. That number would be *e*, the all-important charge on Thomson's electron.

The experiment was far from simple to carry out, however. Millikan found that the water droplets tended to evaporate before any measurements could be made, so he set his graduate student Harvey Fletcher the task of trying the same trick with oil droplets. And that is where the real anarchy began.

When Millikan and Fletcher had refined the technique to the point where it looked as if it would work, Millikan elbowed his student off the project, promising him full credit for some other work. Even Millikan's champion, the Caltech physics professor David Goodstein, admits that this was an act of enormous self-ishness: 'Millikan understood that the measurement of e would establish his reputation, and he wanted the credit for himself.' Fletcher resigned himself to the shortcomings of junior status. 'I did not like this, but I could see no other way out, so I agreed,' he wrote in a memoir that he published only after Millikan's death.

His unfortunate student safely sidelined, Millikan set about using a perfume atomiser and a can of watch oil to create the oil droplets. Some of the droplets would have electrons knocked off as they escaped the atomiser, giving them a positive charge. Others would capture electrons, and emerge negatively charged. Millikan set the electric field across the metal plates at the top and bottom of his apparatus, and watched the droplets rise and fall.

In 1910, at the age of forty-two, he finally published a value for e. It was meant to be his career-defining publication. Eventually, it was – but thanks to scientists working in the German-speaking world, Millikan still had years of difficult and dirty work ahead of him.

The Austrian physicist Felix Ehrenhaft, a strange and intense character himself, quickly disputed Millikan's result. Ehrenhaft had carried out a similar set of experiments and come up with a significantly smaller charge for the electron. In contrast to Millikan's work, Ehrenhaft's experiments seemed to show that electrical charge can be infinitely small. There is no fundamental, minimum unit of charge, Ehrenhaft said; there is no 'electron'. Millikan now had to convince the world that he, not Ehrenhaft, was right. The series of experiments the desperate Millikan then performed were to cast a lasting shadow over his scientific integrity.

The task facing Millikan was to refute Ehrenhaft's claim by showing that the value of charge on the oil droplets was never less than *e*. He was working alone now; Fletcher had gained his doctorate and promptly left for a post somewhere else – possibly *anywhere* else. Millikan took three years to complete the experiment to his satisfaction, and the notebooks he used to record the data show an array of untidy scrawls, enthusiastic exclamations, and rickety rows and columns of numbers. It is clear that Millikan never expected them to come under close scrutiny.

Unfortunately for Millikan's reputation, the historian Gerald Holton retrieved the notebooks from the Caltech archives in 1980. Holton wanted to examine how the clean public face of science compares with the messy process of laboratory work. He was not expecting to spark a controversy that would still be raging decades later.

Much has now been written on the subject of Millikan's honesty. According to the Harvard biologist Richard Lewontin, Millikan 'went out of his way to hide the existence of inconvenient data'. Goodstein, self-appointed defence counsel for Millikan, says that his hero 'certainly did not commit scientific fraud in his seminal work on the charge of the electron'. So where does the truth lie?

The debate hangs on a phrase in Millikan's 1913 paper. In 1910, Millikan had published a value for *e* that is only 0.5 per cent off the value we use today (the error was largely due to his choosing a plausible but mistaken value for the viscosity of air). The 1913 paper was an attempt to refute Ehrenhaft and show that every measurement of electric charge gives a value of *e* or an integer multiple of *e*. In his 1913 paper, Millikan says that his data table *'contains a complete summary of the results obtained on all of the 58 different drops upon which complete series of observations were made'*. The statement is written in italics, as if to give it special weight. The notebooks for the 1913 paper show that Millikan actually took data on 100 oil droplets. The question raised by those who seized upon Holton's analysis was this: did Millikan cherry-pick the data in order to confirm his original result and crush Ehrenhaft underfoot?

He certainly had motive. In Millikan's 1910 paper he had made the 'mistake' of full disclosure. In its pages, he made statements such as, 'Although all of these observations gave values of e within 2 percent of the final mean, the uncertainties of the observations were such that ... I felt obliged to discard them.' Another one is more damning: 'I have discarded one uncertain and unduplicated observation, apparently upon a singly charged drop, which gave a value of the charge on the drop some 30 percent lower than the final value of e.' This admirable honesty about the selection of data points had given Ehrenhaft ammunition that he used enthusiastically in his long feud with Millikan. Perhaps, with the italicised statement, Millikan was making sure that he gave his foe no more.

That would certainly explain something that is otherwise inexplicable. Millikan aborted the experimental run on twenty-five of the droplets in the work reported in the 1913 paper. According to Goodstein, Millikan preferred to use droplets that showed a change in charge, gaining or losing an electron (as he saw it) during the measurement. Millikan may also have judged some droplets to be too small or too large to yield reliable data, Goodstein says. If they were too large, they would fall too rapidly to be reliably observed. Too small, and their fall (and thus the charge result) would be affected by random collisions with air molecules. Goodstein interprets the italicised statement as an assertion that there were only fifty-eight 'complete enough' sets of data.

But that doesn't add up: Goodstein undoes his defence by stating that in order to make the 'too large' or 'too small' distinction, *all* the data would need to have been taken in the first place. Millikan had complete data on seventeen droplets that did not make it into the publication. 'I cannot interpret Millikan's italicized statement as anything other than a lie,' says Caroline Whitbeck, a Professor of Ethics at Case Western Reserve University:

> What would be the point of saying such a thing, let alone putting the statement in italics? Millikan's statement makes sense only as a denial that he has dropped data points. Feeling a need to explain his data selection (which had served him so well), but being unable to fully explain the operation of intuition, Millikan lied.

Millikan certainly did not convince his peers straight away. The arguments with Ehrenhaft rumbled on long enough for Millikan's Nobel Prize to be delayed for three years – it eventually came in 1923. Even after that, things were far from settled. One prominent physicist remarked in 1927 that this 'delicate case' had 'lasted 17 years, and up to now it cannot be claimed that it has been finally decided in favour of one side or the other'.

But here's the point: Millikan was right about the electron and its charge. Few laboratories managed to replicate Ehrenhaft's results, but students now replicate Millikan's results in school laboratories across the world. No one now believes that the fundamental unit of charge is anything other than Millikan's e.

To get his Nobel Prize, Millikan had to play hard and fast with what we might call 'accepted practice'. The science writer George Johnson includes Millikan's work in his book *The Ten Most Beautiful Experiments*, but he is under no illusions about the dark side of Millikan's ambition. 'The beauty here lies with the experiment not the experimenter,' Johnson says.

Scientific anarchy may not be beautiful, but it gets the job done. In 2005, the ethicist Frederick Grinnell made an interesting point in a letter to *Nature*. In basic research, he wrote, intuition is 'an important, and perhaps in the end a researcher's best, guide to distinguishing between data and noise'. By intuition, Grinnell means here the dropping of data points based on a gut feeling that they are inaccurate, just as Millikan did. It's not pretty, and it's not ideal – perhaps not something researchers would be proud to display. But it is what happens.

Grinnell wrote his letter in response to a piece of research about fraud in science. 'Scientists Behaving Badly', by Brian Martinson, Melissa Anderson and Raymond de Vries, caused quite a stir when it was published in *Nature* in June 2005. While the US Office of Research Integrity looks at just three types of fraud – falsification, fabrication and plagiarism – Martinson and his co-authors felt that researchers 'can no longer afford to ignore a wider range of questionable behaviour that threatens the integrity of science'. And so they took a poll, mailing a survey to several thousand scientists to ask what bad things they had done in the previous three years.

Around half the recipients responded. As Martinson pointed out, it is safe to assume that misbehaving scientists are less likely to respond to such a survey than those who consider themselves 'normal', so their results were always likely to be on the conservative side. Nonetheless, the responses were striking. A third of respondents owned up to one or more of the 'top ten' transgressions. These included falsifying data, failing to present data that contradicted their own previous research, changing the design of a study just because funders had asked them to, and stealing someone else's ideas. And all this in just a three-year period.

As it turned out, scientists in mid-career – as Robert Millikan was – are more likely to commit such sins. Those who feel aggrieved at the way the funding process works are also more likely to misbehave. To add to the issues, status matters in scientific misbehaviour. 'Star scientists' are more likely to commit scientific sins, but less likely to get caught, than the average scientist.

Albert Einstein's name, which is synonymous with genius, crops up in a few chapters of this book. But it is perhaps most interesting to look more closely at him as a scientist behaving badly. If Einstein were forced to fill out Brian Martinson's 'Scientists Behaving Badly' survey honestly, he would have to admit to committing five of the sixteen misdemeanours listed. When we allow for the fact that three of them cannot be applied to his particular field, that gives us a hit rate of just over one-third. Einstein provides a perfect example of the character that will produce great science and think nothing of the misdemeanours that such breakthroughs demand.

Einstein's track record would set off alarm bells in university personnel departments today. The close attention such a genius receives has exposed him as a womaniser who made shameless passes at his mistress's daughter. When confronted by both women, he shrugged and asked them to decide which of them he should marry once he had divorced his wife, Mileva. In his divorce settlement he arrogantly promised to give Mileva the money from an as yet unearned Nobel Prize. When the prize money did come in, he gave her only half. He let his alimony payments dwindle to intermittency. He made his university pay him a full-salary pension on retirement, threatening to use his fame against them should they refuse. He hid money from the taxman, and he cut off his schizophrenic son, leaving him to die a 'third class' patient in a mental institution.

None of this proves anything about Einstein's scientific integrity, but it would be naive to think that a scientist's humanity is neatly compartmentalised; personal and professional lives are unlikely to be so neatly separated that character traits obvious in one will have no bearing on conduct in the other.

Was Einstein fundamentally dishonest, a cheat in science as well as love? No. But there are plenty of shady moments in his professional life. Putting this heroic scientist into the best possible light, it is clear that he was an enthusiastic and gifted thinker who, in his determination to understand the universe and make it understood to others, considered the accepted practices of science as guidelines rather than laws carved in stone. He was well aware that convention demanded the game should be played a certain way, but sometimes it suited him to defy convention. He was certainly not above picking the data that worked best with his theories, for example.

In early 1915, a little cherry-picking must have seemed a very minor scientific misdemeanour. On the outskirts of Warsaw, the German army was busy putting the creations of German chemists to the test. Xylyl bromide – better known as tear gas – was a disappointment when it was first used: it was cold in Poland in January, and the gas froze instead of dispersing. When the army switched to using lethal gas shells, it was a different story. At Ypres, chlorine gas killed 6,000 Allied soldiers in just ten minutes.

Meanwhile, Einstein, who had bravely registered his objections to the war at considerable risk to his own status and safety, was conducting his own private battle. His effort to generalise his special theory of relativity, to describe how the presence of energy and mass affects the fabric of the universe, was getting nowhere. To relieve his frustration, he wandered into a Berlin laboratory and began to tinker with some iron magnets.

The aetherian explanations of electrical disturbance held no appeal for Einstein: he believed in Thomson's electron. Furthermore, he suspected that magnetism resulted from the circulatory motion of electrons within atoms of iron. As a diversion, he decided to put the matter to the test.

With the help of a colleague, he suspended an unmagnetised iron rod from a glass fibre, then used a magnet to change the magnetism of the iron. If his suspicions were correct, changing the magnetism would change the amount of circulatory motion within the rod. By the law of conservation of angular momentum, it is impossible to do that without causing some opposing, compensating motion. The iron rod would be forced to rotate in the opposite direction to the electrons in order to preserve the angular momentum. And that is exactly what Einstein found.

His theory predicted that a particular amount of magnetism would induce a particular amount of motion. The exact ratio of magnetism to motion, called the gyromagnetic ratio, would be 1. His experiment put the value at 1.02, close enough 'to silence any doubt about the correctness of the theory', as he told the German Physical Society in his report. 'A wonderful experiment!' he wrote to Michele Besso, a friend and former colleague at the patent office in Bern. 'A pity you didn't see it.'

When others tried to replicate the experiment, however, things weren't so wonderful. After six years of testing, the gyromagnetic ratio was found to be 2. Einstein, guided by his own (erroneous) theory, continued to refuse to believe that it was anything other than 1. Many years later, Einstein's collaborator in the experiment, the Dutch physicist Johannes de Haas, admitted that they had done the experiment twice, obtaining values of 1.02 and 1.45. Einstein had picked and published the value that matched his theory.

It is hardly a major crime. But Einstein's little misdemeanour does tell us two things. First, cherry-picking is rarely punished. It is just what you do to get science done. Sometimes, as with Millikan, it works, and history paints you a hero. Sometimes, as with Einstein and the gyromagnetic ratio, it doesn't, and history shrugs

– either because it doesn't matter much, or because you are found out only when others get the right answer. The catcalls that might be aimed at you are drowned out by the applause directed at those who have succeeded.

The second insight from that episode is perhaps the more interesting. Einstein was entirely cavalier about the 'sacred' processes of science. And so, to some degree, are all scientists. Einstein once advised that if you want to know how theoretical physics gets done, the last person you should ask is a theorist. 'I advise you to stick closely to one principle: don't listen to their words, fix your attention on their deeds,' he said.

He was aware that such an attitude was not how science should be presented, and was fond of making public statements such as the famous 'no amount of experimentation could ever prove me right, while a single experiment could prove me wrong'. Such fine words are all very well, but the fact remains that Einstein refused to accept the gyromagnetic ratio as anything other than the value offered by his theory. He had a similar mindset when it came to the theory of relativity. It would always be correct in his eyes, even if experiments proved it not to be so. 'I would feel sorry for the Dear Lord,' he once told a student. 'The theory is right anyway.'

For theorists, this is an entirely defensible stance. Paul Dirac's take on the problem of theory vs experiment is similar: 'If there is not complete agreement between the results of one's work and experiment, one should not allow oneself to be too discouraged,' he said. The one exception to the acceptability of such attitudes is when theorists deliberately fudge their theory. Fortunately for Einstein, when he committed *that* sin, the Dear Lord was kind.

In general, theorists are thought to be immune from the worst scientific misdemeanour, the one that tops Martinson's table.

Falsifying or 'cooking' research data – 'counterfeiting the coin of science', as David Goodstein has put it – is generally thought to be impossible for those who deal only in ideas. But that is simply not the case. Constructing a mathematical theory is not unlike performing an experiment. Every step requires attention to detail; one slip renders the whole endeavour invalid. You must watch carefully for unwarranted assumptions, for example: mathematical models are developed to deal with particular situations, and what applies in one scenario will not necessarily work for another. Just because a formula applies in one context, such as when moving at the speed of a train, that doesn't mean that it applies in another – moving at something close to the speed of light. Einstein, ever the anarchist, refused to let such inconvenient details get in the way of a good idea.

In 1905, Einstein's intuition told him something 'jolly and beguiling': that the mass of a body will change if it emits a pulse of light. This crystallised in his mind as the famous $E = mc^2$ equation: the energy lost as the light pulse is equal to the change in mass multiplied by the square of the speed of light. But he never managed to prove it.

His first attempt – the 1905 paper published in September's *Annalen der Physik* – contained a mistake. Einstein used a formula that applied only to slow-moving bodies. The description of fast-moving emitters of light required an entirely different approach. According to the physicist Hans Ohanian, this error 'is the sort of thing every amateur mathematician knows to watch out for', but Einstein didn't bother with it. Generously, Ohanian suggests that Einstein's mind must still have been exhausted from his work on special relativity, which had been finished only a couple of months before. But over the next forty-one years, Einstein made eight attempts at a proof of $E = mc^2$. Not once did he manage it without inserting a fudge.

Take Einstein's 1912 'proof', for example. The approach he took was borrowed (without acknowledgement) from work done by another physicist, Max von Laue. In the process of trying to make it his own, Einstein had to admit to making a nonsensical assumption. One footnote reads, 'To be sure, this is not rigorous.' Mitigating his fudge, he suggests that the idea that the assumption won't work out in his favour 'seems so artificial that we will not dwell on this possibility at all'. This is not an attempt at fraud, or even a glossing-over of inconvenient truths. It is more like a mind-trick, an illusionist using the power of suggestion. If it was Newton's style that Einstein copied when he fudged the mathematics, here he was borrowing Galileo's tactic of bullying others into asking no questions.

The final attempt to prove $E = mc^2$ came in 1934, when Einstein presented a 'repaired' proof of the equation to a gathering of scientists. A *New York Times* reporter was in attendance, and made it front page news. The writer gushed about Einstein's talk: it was like 'watching a Beethoven making the final draft of his Ninth Symphony'. Four hundred American scientists were given 'the treat of watching him remodel his universe. A piece of chalk was his only tool.' But the proof was still wrong – for the same reason that his first proof was wrong. The error had been pointed out years before by no less an authority than Max Planck, the creator of quantum theory, but Einstein had either not noticed or decided to ignore Planck's advice.

It wasn't as big a deal as the *New York Times* made out, anyway. Nobody in the know had been surprised by the equation, even back in 1905: the relation was known to exist for electrical energy, if not light. And by 1934 several mathematicians had already published ironclad proofs that stood in stark contrast to Einstein's botched attempts. By this time, though, Einstein had appropriated the equation as his own. He dismissed attempts to set the record straight with contempt or aggressive assertions of his 'priority'. Not until 1949, when he published an autobiography, was there an inkling that Einstein was prepared to back down. Though the volume refers to all of his many genuine contributions to physics, $E = mc^2$ is nowhere to be found.

Moving down to the ninth item on Martinson's list, we find another of Einstein's misdemeanours. He was guilty of 'overlooking others' use of flawed data or questionable interpretation of data'. But are we really surprised? After all, Arthur Eddington's data supported Einstein's theory, and we have already seen how well disposed Einstein would have been to that.

Questions about Eddington's use of data have been raised many times. What has received less attention is his most essential motivation: not to prove Einstein's beautiful theory right, but to bring an end to hostilities between nations.

Arthur Eddington was a Quaker. Though the Quakers are often seen today as a mild-mannered group, welcoming everyone and insisting on nothing, that was certainly not the case at the turn of the nineteenth century. Eddington's values were forming in the years when the Quakers were radicals. They rejected traditional Christian views, were happy to rely on their brains rather than the Scriptures for guidance, and most of all were eager to see good in all humankind, regardless of colour or creed. Quakers had begun the campaign for the abolition of slavery in the seventeenth century, decades before William Wilberforce lent his weight to the movement.

When the First World War broke out, then, Eddington was ready to fight – but only for the cause of pacifism. His active, radical Quakerism spurred him to look for ways in which he might wage war on the division between nations. He had made clear his feelings that the war should not affect the working relations between scientists on opposite sides of the conflict. In April 1918, when he was called up to active service in the British army, an opportunity to further that cause presented itself.

Eddington refused to join up, becoming a conscientious objector. This provoked a long series of hearings at which various influential colleagues tried to argue that because of his position as Director of the Cambridge Observatory he should be excused military service. Eddington, no doubt to the utter exasperation of his influential colleagues, undercut these objections. His scientific importance was not the reason he sought exemption, he said. 'My objection to war is based on religious grounds,' he told the Cambridge tribunal board that heard his case. 'I cannot believe that God is calling me to go out and slaughter men.'

It was a dangerous tactic. The much depleted British army was desperate for more recruits. Conscientious objection was no longer considered a valid reason for avoiding service, and Quaker colleagues, such as Ebenezer Cunningham, a mathematician at St John's College, had just had their objections to the call-up swept aside. Objectors, the object of scorn and persecution among British soldiers and the British public alike, were being conscripted despite their protests, and being assigned to minesweeping and other perilous tasks.

Eddington was saved from this fate only when the Astronomer Royal, Frank Dyson – perhaps understanding Eddington better than most – gave him a dignified way out. Dyson had been intrigued by Einstein's general theory of relativity since its announcement in 1915. He was sceptical about its claims, and had sought ways to prove or disprove it. The only way forward seemed to be to find a way to test its prediction that the presence of mass bends space. The bending of space, Einstein said, meant that light would not always travel in a straight line. If he was right, the light from distant stars would follow a curved path as it passed near the

Sun, for example. That curve would make those stars appear to be slightly displaced in the sky.

It sounds an easy thing to test, but there are two complications. The first is that Newton's physics also says that light's path will bend in the presence of a large gravitational field. The Newtonian effect is about half as much the effect that relativity predicts. The second complication is that there is an obvious difficulty in looking at stars that are almost aligned with the Sun. Any telescope that can pick out the starlight would blind the astronomer with sunlight. The only solution was to perform the observation during a total eclipse of the Sun.

Dyson had already looked at photographic plates exposed during previous eclipses, but he found nothing that could prove or disprove Einstein's theory. He had worked out, however, that a total eclipse due in 1919 would provide the required data – as long as the scientists were prepared. For any eclipse, totality occurs only along a narrow strip of the Earth's surface. The result is that only a few observing sites offer the necessary darkness. To make the observations, Dyson decided, a team of astronomers would need to embark on a long and arduous expedition to Principe, an island off the coast of West Africa.

Dyson told the Military Service Tribunal hearing Eddington's case that Eddington's work ranked alongside that of Darwin, and then reminded them that the pre-eminence of British science was in question. There was, he said, 'a widely spread but erroneous notion that the most important scientific researches are carried out in Germany'. If Britain's astronomers were given enough time to prepare for and undertake the expedition, the Principe observations would put that matter to rest and restore the pride of Britain. 'Prof. Eddington is particularly qualified to make these observations, and I hope the Tribunal will permit that important work to be continued.'

The ruse worked – on both parties. The Tribunal allowed Eddington to carry on with his research, and Eddington, to everyone's relief, accepted the exemption. As Matthew Stanley, one of Eddington's biographers, has put it, 'it was an opportunity for him to bring a peace-loving, insightful German to prominence in both science and society'.

In other words, Eddington already believed that Einstein was right, and he was quite prepared to accept a God-given opportunity to prove it. The expedition to Principe was a chance to bring peace on Earth. In a statement Eddington made years later, he pointed out that his confirmation of Einstein's theory was 'not without international significance', because by 'standing foremost in testing and ultimately verifying the "enemy" theory, our National observatory kept alive the finest traditions of science; and the lesson is perhaps still needed in the world today'.

Having established that Eddington had motive – perhaps the best of motives – for ensuring that Einstein's theory was validated, we can now look with eyes wide open at what he did to achieve it.

If the Divine Hand did send Eddington to Principe, it did nothing to help with the observations. The expedition was faced with a barrage of rainstorms, and its members had to build waterproof shelters for the equipment. Thanks to the island's lively insect population, everyone had to work dosed up on quinine and sheltering under mosquito nets. At night, monkeys would emerge from the forest and, fascinated by the telescopes, clamber over them and interfere with the settings. Enraged, the scientists joined the technicians in hunting down and killing these intruders.

And then, on the morning of eclipse, the heavens opened again. Though the rain stopped two hours before totality, a blanket of grey cloud still covered the entire sky. Just in time it cleared a little, and Eddington was able to take some pictures of the Sun through his telescope, but the cloud 'interfered very much with the star-images'. Not surprisingly, the photographs were a let-down. During the eclipse, Eddington captured sixteen images on photographic plates. A week later he had managed to develop twelve of them, but only two were useable: the majority 'show practically no stars', he said. 'It is very disappointing.'

So disappointing, in fact, that Eddington decided to abandon his original notions of how to calculate the positions of the stars from the measurements on the plates. Instead, he formulated a new technique for the purpose. This new method involved a few assumptions and Einstein's own numbers for what the displace-ment should be. Perhaps unsurprisingly, Eddington achieved a rather pleasing result: 'the one good plate that I measured gave a result agreeing with Einstein and I think I have got a little confir-mation from a second plate'.

The result was a displacement of 1.61 arc-seconds. Einstein's theory – as Eddington knew – predicted 1.75 arc-seconds. New-ton's theory, the status quo, said that the displacement would be 0.8 arc-seconds. These are tiny shifts, roughly equivalent to the diameter of the smallest coin in your pocket seen from a mile away. But Eddington was happy to declare Einstein the winner. As he wrote the following year, 'Although the material was very meagre compared with what had been hoped for, the writer (who it must be admitted was not altogether unbiassed) believed it convincing.'

But Eddington's was not the only expedition to test general relativity at the 1919 eclipse. The Astronomer Royal had also sent a team to Sobral in the north-east of Brazil. They were equipped with an astrographic telescope, just like Eddington's, and had enjoyed fine, clear weather, which had allowed them to come away with plenty of photographs. However, as it turned out, fine, clear weather brought its own disadvantage: the Brazilian heat had distorted a mirror used to focus starlight into the main telescope. As a result, the images obtained were slightly fuzzy. A note written on 30 May, after four of the plates had been developed, admits that 'It seems doubtful whether much can be got from these plates.'

But a value was derived nonetheless: a deflection of 0.9 arcseconds. This was far too low to confirm Einstein's theory, and very close to the standard Newtonian explanation for the passage of light through the universe. Fortunately for Eddington, the Brazilian expedition had taken along another, smaller telescope. When the images obtained with this instrument were analysed, a deflection of 1.98 arc-seconds was found.

In their book on science and its methods, *The Golem*, Harry Collins and Trevor Pinch demonstrate that a modern analysis, one that takes into account all the experimental results, tells us that no conclusion can be drawn from the 1919 British eclipse data. The eight 'good' Sobral plates yield a displacement of just over 1.7 arc-seconds; the two 'poor' Principe plates give a value between 0.9 and 2.3 arc-seconds. The mean of the 'poor' Sobral plates put an upper limit of 1.6 arc-seconds on the displacement of the stars.

By November, though, Eddington had decided which of the data were the most valuable: his own two plates. They were the fuzziest of all, and the values derived from them had been obtained by using a formula that included Einstein's own prediction for the result. Nevertheless, J.J. Thomson, now the President of the Royal Society, ruled the evidence admissible.

Perhaps Thomson was particularly sensitive to accusations over questionable data. The debate between Millikan and Ehrenhaft over Thomson's electron was still rumbling, after all. And he would have been familiar with that scientist's sense of just 'knowing' when something is right, even when a truly satisfactory proof remains elusive. So, despite the mutterings anyone might have heard, Einstein had been proved right. 'It is difficult for the audience to weigh fully the meaning of the figures that have been put before us,' Thomson told the assembly at the Royal Society, 'but the Astronomer Royal and Professor Eddington have studied the material carefully, and they regard the evidence as decisively in favour of [Einstein's] value for the displacement.'

Thomson's announcement obviously didn't have its intended impact. The Nobel committee excluded relativity from Einstein's 1921 Nobel Prize in Physics (awarded, for complex reasons, in 1922): the last sentence of the letter sent to Einstein states that the prize was being awarded 'in consideration of your work on theoretical physics and in particular for your discovery of the law of the photoelectric effect, but without taking into account the value which will be accorded your relativity and gravitation theories after these are confirmed in the future'.

That careful phrasing must have been a knife in the back to Eddington, but it seems fair. Even in 1962, when a team of astronomers tried to reproduce Eddington's findings during an eclipse, they failed – despite having better equipment. They concluded that it just couldn't be done. Small wonder, then, that many of Eddington's contemporaries remained sceptical about the 'evidence' for Einstein's theory. In 1923, one commentator summed up the situation:

> Professor Eddington was inclined to assign considerable weight to the African determination, but, as the few images on his small number of astrographic plates were not so good as those on the astrographic plates secured in Brazil, and the results from the latter were given almost negligible weight, the logic of the situation does not seem entirely clear.

So what did Einstein make of Eddington's work? Did he suggest we await less controversial confirmations of relativity? Of course not. Einstein was the anarchist-in-chief.

Einstein 'knew' that he was right before the confirmation. His friend Heinrich Zangger, a professor of pathology at Zurich, heard about Eddington's results and wrote to Einstein: 'Your confidence … that light would have to go bent around the Sun … is for me a tremendous psychological experience. You were so certain, that your certainty had an overwhelming effect.'

It may even be that Einstein never bothered to find out exactly what Eddington had seen. Einstein was certainly cavalier with the facts about Eddington's work: writing to Max Planck, he declared that 'the precise measurements of the plates yielded exactly the theoretical value for the deflection of light'. This was, as we have seen, not true. But it is not clear that Einstein was deliberately misleading Planck. Einstein was simply not terribly interested in data.

Genius that he was, this was an attitude that served him well. In 1905, for example, his ideas on special relativity did not fit with the available data on how electric and magnetic fields deflected beams of charged particles; instead, it confirmed a rival theory. But Einstein would have none of it. He declared that the competing theories were inadequate when faced with other kinds of experimental results and 'should be ascribed a rather small probability [of being right]'. He was absolutely correct: a little while later, more accurate measurements showed that special relativity was the better theory.

Einstein once told the German theoretical physicist Werner Heisenberg that it would be 'quite wrong to try founding a theory on observable magnitudes alone. In reality the very opposite happens. It is the theory which decides what we can observe.' That thinking seems to be behind this statement of Eddington's:

> Observation and theory get on best when they are mixed together, both helping one another in the pursuit of truth.

It is a good rule not to put overmuch confidence in a theory until it has been confirmed by observation. I hope I shall not shock the experimental physicists too much if I add that it is also a good rule not to put overmuch confidence in the observational results that are put forward until they have been confirmed by theory.

In science, anything goes. Or does it? Perhaps, by focusing on a few celebrated scientists, I have been guilty of doing my own cherry-picking. To claim from these cases that science is anarchic, and that it does not enslave itself to the results of experiments, is all very well. But does the claim stand up when applied to everyday science? Yes it does.

Perhaps the most compelling evidence for modern, everyday scientific anarchy came in 2006, from grass-roots interviews carried out by the team of researchers responsible for the 'Scientists Behaving Badly' survey of 2005, discussed at the beginning of this chapter. The results were so clear-cut that Raymond De Vries and his colleagues called their 2006 paper 'Normal Misbehavior'. What's more, they saw the common misbehaviours as having 'a useful and irreplaceable role' in science.

The conversations that led to this startling conclusion were held with fifty-one scientists who were roughly one-third to halfway up the career ladder: assistant professors and postdoctoral fellows at public and private 'major research universities'. The interviewees' statements make entertaining and enlightening reading. Here's one:

Okay, you got the expected results three times on week one on the same preparation, and then you say, oh, great. And you go to publish it and the reviewer comes back and says, 'I want a clearer picture,' and you go and you redo it – guess what, you can't replicate your own results … Do

you go ahead and try to take that to a different journal …
or do you stop the publication altogether because you can't
duplicate your own results? … Was it false? Well, no, it
wasn't false one week, but maybe I can't replicate it myself
… there are a lot of choices that are gray choices … They're
not really falsification.

And another:

[T]here was one real famous episode in our field [where]
it was clear that some of the results had just been thrown
out … [When they were] queried [the researchers] … said,
'Well we have been doing this for 20 years, we know when
we've got a spurious result …'

There is another compelling indication that such attitudes are
endemic: you can't erode them by education or policing. I have
already mentioned that the proportion of 'questionable data' published in *Nature Cell Biology* remained unaffected by the introduction of a data-screening process. The introduction of ethics classes
for researchers at the University of Texas had similarly negligible
effects. In fact, in a bizarre twist, a 1996 study of the effect of ethics education found that students became more likely, not less, to
commit certain types of misdemeanour.

Now to strike the final nail into the coffin of the idea that scientists abhor and avoid misconduct. It is a clear and established
fact that even those convicted of the worst crimes can return to
the fold.

In 2008, University of Pennsylvania researchers Barbara Redman and Jon F. Merz published a rather remarkable piece of
work. They tracked down forty-three established, high-flying
researchers who had been found guilty of major misdemeanours
such as falsifying results. Through literature searches, telephone

interviews and dogged detective work, Redman and Merz established that, just a few years after their conviction, many of those offenders were back at the coalface, collaborating with colleagues and publishing scientific papers.

Reading the paper, you get the impression that Redman and Merz are somewhat shocked by their findings. They report that 'the picture of the consequences painted by our interviews, which shows both the hardship of punishment and the chance for redemption, is perhaps more positive than it should be.'

If they are shocked, it is only because they failed to realise that, to a large degree, scientists gloss over misconduct. Here is what Richard Smith, former editor of the *British Medical Journal*, had to say about the problem:

> Most cases are probably not publicized. They are simply not recognized, covered up altogether; or the guilty researcher is urged to retrain, move to another institution, or retire from research. I have spoken perhaps a dozen times on research misconduct in several countries and often to audiences where people come from many countries. I usually ask the members of these audiences how many know of a case of misconduct. (I consciously do not offer a definition.) Usually half to two-thirds of the audience put up their hands. I then ask whether those cases were fully investigated, people punished if necessary, lessons learnt, and the published record corrected. Hardly any hands go up.

In De Vries's words, such statistics result from scientists' acknowledgement of 'the ambiguities and everyday demands of scientific research'. In other words, everyone knows it's the only way to get the job done.

Science is a quest to convince yourself and others of something you only guess to be true. That is a hard task, and requires tenacity and ingenuity – and, occasionally, questionable tactics. Major frauds, such as fabricating or copying results, are unlikely to go undetected, but scientists also know that results so gained are unlikely to satisfy the inner craving that drives them. That is why only 1 in 300 perform such acts. The minor misdemeanours, on the other hand – the cherry-picking or the questionable methods of analysis, are weapons with which to swipe at the irritating but inescapable ambiguities without risking any dishonour or self-doubt.

As long as things turn out OK, those who bend the rules in this way are almost always forgiven. The science writer Simon Singh, discussing accusations of cherry-picking in Eddington's work, concedes that Eddington may have 'subconsciously minimised his errors in order to get the right result'. But Singh waves this indiscretion aside: 'Regardless of whether or not this was the case, Eddington's result was hailed as a wondrous piece of science.' Goodstein's defence of Millikan includes a similar observation: 'it is worth remembering that history has vindicated Millikan in that his result is still regarded as correct', he says. Heredity pioneer Gregor Mendel's data are suspiciously clean, and that has been largely overlooked because his hypotheses turned out to be correct.

The dictates of science say that it is impossible for scientists to prove themselves right without carrying out experiments. Unfortunately, running useful experiments according to the accepted scientific method often proves impossible too. So results get fudged. Years, decades or even centuries later, with all the ambiguities of their experimental methods exposed, we still see these people as great scientists. And rightly so.

If their convictions and intuitions had been wrong, these

scientists would have disappeared from history. It is the intuitive understanding, the gut feeling about what the answer should be, that marks out the greatest scientists. Whether they fudge their data or not is actually immaterial. Simon Westfall's study of Newton led him to convict Newton of deliberate fraud, and yet, when all was done, he remained in awe: 'He has become for me wholly other, one of the tiny handful of supreme geniuses who have shaped the categories of the human intellect.'

Faced with all this, the historian of science Stephen J. Brush has gone so far as to ask whether his subject should be X-rated. The myth of the scientist, as a rational, open-minded investigator who proceeds methodically, is grounded in the outcome of controlled experiments and searches objectively for the truth, is a useful one, he says. If young scientists were to find out what really happened in the history of their subject, it might 'do violence to the professional ideal and public image of scientists'.

But surely the truth is to be celebrated. This is real science, as done by very human beings. And, as we have seen, this is *good* science. As time goes on, further experiments, using better technology or new ideas, exonerate those who were brave enough to be right without the full support of the data. This is the way progress has happened. Perhaps it is the only way it *can* happen when established – but wrong – ideas refuse to be uprooted by any other means. If sweeping aside one tide per day, declaring a problematic photographic plate void, or disregarding an obstinately stable oil droplet is the price we pay for beating a quicker path to the truth, then so be it. The future must be brought into the present by any means necessary. And, as we shall see in the next chapter, that includes finding ways to deal with situations where there is no decisive evidence at all.

3

MASTERS OF ILLUSION

Evidence isn't everything

'Why are you killing off all the women, stealing our faeces from the latrines to perform sorcery?' In 1962, Shirley Lindenbaum, an anthropologist living with the Fore tribe in the lush eastern highlands of New Guinea, witnessed an extraordinary spectacle. The Fore women had gathered the entire tribe together to accuse their men of witchcraft and murder.

'We women give birth to you men,' the spokeswoman continued. 'Try to find one man who is pregnant now and show him to us. Or go and search the old burial grounds and bring us the skull or bones of one man we women have killed. You won't be able to find any. You men are trying to wipe us out.'

The complaint was not without justification. The tribeswomen, and many of their children, were being struck down by a mystery illness. In ten years, more than a thousand women and children had fallen victim to kuru, the shaking disease. The Fore men, on the other hand, were almost untouched. By the early 1960s the impact of the disease was such that there were three Fore men for every woman. The birth rate fell almost as fast as the marriageable age. The shortage of women meant that girls were now being married off almost as soon as they entered puberty.

The first symptoms of kuru are an unsteadiness on the feet, slurring speech, tremors and shivers. Later there are outbursts of laughter, and then the muscles start to spasm and jerk. Victims fall into depression, lose the ability to walk and become doubly incontinent. Death comes as a merciful release.

Virologist Carleton Gajdusek was the first doctor to come to the Fore's aid. He arrived from Melbourne in 1957, but for years he made little headway. His photographs of the tribespeople are heartbreaking: the women and children are walking with sticks, or beyond help and needing to be carried. These are some of the captions he wrote for the photographs:

> A girl, about 8 years old, who was no longer able to speak, but who was still alert and intelligent … Four preadolescent children, totally incapacitated … none had been ill for over six months, and all died within a few months of the time of photography … The youngest patient with kuru, from Mage village, North Fore, who self-diagnosed the insidious onset of clumsiness in his gait as kuru at 4 years of age, and died at 5 years of age, several years before his mother developed kuru herself.

Gajdusek understood the Fore's reasoning that sorcery must be to blame. He came to respect the local traditions, even accepting that their magical reasoning had positive repercussions:

Patients know they are to die; they have observed the terminal incapacitating stages of the disease in others, and, yet, discuss the matter of their advancing illness freely and without apparent anxiety. They will laugh at their own stumbling gait and falls, their clumsiness, inability to get food into their mouths, and their exaggerated involuntary movements, and their kinsmen will join them. The family members live with the dying patient … parents sleep with their kuru-incapacitated child cuddled to them, and a husband will patiently lie beside his terminal uncommunicative, foul-smelling spouse … the emotionalism and euphoria of kuru is supplemented by a security engendered from certain knowledge that one is accepted by his villagers as an unfortunate victim of kuru sorcery, whom they will not desert before death claims him. The vengeful search for the offending sorcerer, which is often the primary concern of the patient's kinsmen, is a source of further emotional support.

Western medicine, Gajdusek noted, had nothing to offer. His letters home convey a sense of uselessness:

[T]hey know damn well that we do nothing for the disease, but prolong its misery by supportive measures, and they are anxious to return to their technique of starvation and neglect in darkness, which ends in a speedy exodus once the illness is truly incapacitating.

Despite the depressing futility of his efforts to care for the Fore, Gajdusek continued to study kuru. It was frustrating work: the disease seemed to strike from nowhere. He had an idea that it might be genetic – the cases seemed to be grouped in families, after all – but even that was proved wrong. Many members of the family groups he had identified turned out to be genetically unrelated. Eventually, however, Gajdusek did find the cause: cannibalism.

Keen to ingest the wisdom of their forebears, the Fore in their funeral rituals ate their dead ancestors. The ritual cannibalism began with the women stripping muscle from the body to give to the male members of the family. Having handed over the choice cuts, the women fed themselves, their children and elderly relatives on the internal organs, including – especially, in fact – the brain.

In an experiment that was to win him a Nobel Prize, Gajdusek showed that injecting chimpanzees with material from the mashed-up brain of a kuru victim brought out kuru-like symptoms in the chimps. Kuru, he said, was infectious: the result of an as yet unidentified 'slow virus' that lived in the organs the women and children ate, but not in the muscle that the men consumed.

Gajdusek's breakthrough paper was published in 1966. He and his colleague Joe Gibbs spent the next few years testing whether dozens of other human neurological diseases could be transmitted in the same way. Only one gave a positive result: Creutzfeldt–Jakob disease, or CJD. That result was published in 1968, the same year that Stanley Prusiner, the anarchist who stands at the centre of this chapter, received his MD.

Stanley Prusiner first encountered CJD when he was working as a resident in neurology at the University of California, San Francisco. It was 1972, and one of his patients was suffering a slow and agonising death. The typical symptoms of CJD are confusion, blindness, jerking spasms and an inability to communicate. These patients, we can safely assume, die lonely, frightened and in darkness. Fascinated and appalled, Prusiner began to read the scientific literature on the disease. From this he learned that his patient had succumbed to a mysterious slow-acting virus that no one had ever been able to isolate or identify. It was a mystery that would define the rest of his career.

Kuru and CJD are now known to be two of a select group of diseases that includes bovine spongiform encephalopathy (BSE, also known as mad cow disease) and the sheep and goat disease scrapie, so called because its symptoms include a desperate itching that causes the animals to scrape off patches of their wool or hair. Prusiner began his research with a programme of reading about scrapie, the disease in the group that had been studied the most. Almost immediately he came across what he later called an 'astonishing' report by Tikvah Alper, then at the Hammersmith Hospital in London. In 1967, Alper and her colleagues had shown that the mysterious virus might not be a virus at all.

Viruses multiply by coercing a host cell into reproducing the virus's genes. But genes are composed of chains of molecules known as nucleic acids, and these are fragile. If you subject them to ultraviolet or ionising radiation, the chains break up. For organisms like us, that often results in life-threatening cancer. For viruses it is even more catastrophic: radiation kills them. So when Alper's team took the brains of scrapie-infected sheep and irradiated them, they fully expected the viruses to be killed and the brain material to be rendered safe.

It was not. Even after irradiation, the brain material remained infectious. It wasn't a difficult step to suggest that, whatever the infectious agent was, it didn't contain the nucleic acids that make up genetic material. That is, it wasn't a virus.

Alper had already thrown some cold water on the virus idea. She had weighed the minimum amount of material that would cause an infection, and shown that its molecular weight was too low for it to be a virus or a bacterium. What's more, no one had ever found the nucleic acids. So, if not a virus, what was the infectious agent? The English mathematician John Stanley Griffith had come up with an answer to that in a September 1967 *Nature* article. He suggested that a protein could be responsible.

Proteins are long strings of acid molecules that play specific roles within the body after folding themselves into particular shapes. Haemoglobin is a protein that carries oxygen through the body, for example, and insulin is a protein that signals the presence of excess sugar that needs storing. But for all the impressive cleverness of proteins, Griffith's suggestion seemed, at first glance, ridiculous. To cause an infection, you need an agent that can reproduce itself – and proteins can't do that.

In orthodox molecular biology, the instructions for replication are encoded in nucleic acids. Viruses contain nucleic acids. So do bacteria. That's why they can reproduce, and that's why the mainstream view was that scrapie must be caused by a slow-acting virus. A protein, however, is the *product* of the instructions encoded in nucleic acids. There is, on the face of it, no way that a protein can be a creative entity.

Griffith knew that, but as a mathematician he wasn't constrained by the norms of molecular biology. He put forward three ways in which a protein might conceivably replicate itself. 'There is no reason to fear that the existence of a protein agent would cause the whole theoretical structure of molecular biology to come tumbling down,' he said. And he suggested that scrapie infections could quite conceivably be caused by a protein that the animal is 'genetically equipped to make'. Perhaps, he said, the animal wouldn't normally make this protein, at least not in a form that folds into this shape. But, passed from another animal, the protein might stimulate the animal's natural proteins to take on another, dangerous shape.

Prusiner was intrigued, and decided to look more closely at what the infectious agent might be. By 1974, he had secured the funds to set up a laboratory that processed scrapie-infected brains to extract the infectious material in as pure a form as possible. Equipped with what he terms 'the optimism of youth' and a 'cocky' nature, he succeeded where others had failed. By 1982, he was able to report that the concentrated infection agent for scrapie consisted almost entirely of protein. It appeared to contain no nucleic acids – and thus no genetic material. It was not, as most people thought, a virus. It was, Prusiner said, a protein, just as Griffith had suggested it might be.

Prusiner made the claim in a paper published in the journal *Science* in 1982. It was meant to be a review of scientists' current understanding of scrapie. Instead, Prusiner used the commission from the journal as a platform from which he could announce an entirely new discovery. Biology says that there are only two types of infectious agent: the bacterium and the virus. Stanley Prusiner now introduced a third: the 'prion', a clever contraction of 'proteinaceous infectious agent'.

He pronounces it *pree-on* – though I was interested to note during a telephone conversation that Laura Manuelidis, the head of neuropathology at Yale University, pronounces it *pry-on*. I'm sure it doesn't matter to her; she still doesn't believe that prions are real, and no one else can be sure whether to believe that prions exist. As a February 2010 article in *Science* pointed out, 'three decades of investigation have yielded no direct experimental proof' that the cause of these infections is exclusively down to a protein. Not that this held back Prusiner's Nobel Prize.

'At every crossroads on the road that leads to the future, tradition has placed against us ten thousand men to guard the past.' That was how, in 1992, Prusiner chose to preface a collection of conference papers about his work. The bitter tone was intentional: his colleagues and peers had been standing against him for a decade, ever since that *Science* paper. Prusiner claims that the acceptance of his research represents 'a triumph of scientific process over prejudice'. So where did this prejudice-driven rejection come from? From those who believe that evidence is the most important thing in science.

Byron Caughey, laboratory chief at the US National Institutes of Health's Rocky Mountain Laboratory in Montana, describes Prusiner as doing 'horrendous things, and just running roughshod over his field'. Other scientists were 'more careful', he says: they 'wanted to stick close to the evidence'. The press soon got wind of the developing ruckus. They cast Prusiner as 'prickly and aggressive', a heretic, a charlatan, so consumed by his own ideas that he had abandoned objectivity. Why? Because prions weren't – and still aren't – the only game in town.

The fact that no one had found any nucleic acids associated with the infectious agent didn't mean that they weren't there. In 1982, shortly after *Science* published Prusiner's paper, Richard Kimberlin of the Animal Research Centre in Edinburgh countered it in *Nature*. It was entirely possible, he said, that the infectious agent was an information-carrying molecule, the 'virino', most probably made of nucleic acid but contained in a protein coating. Whereas a virus encodes the instructions for making its proteins and nucleic acids, a virino would encode only the instructions to make the nucleic acids. The genome of the host organism would specify what proteins would be made in the event of infection.

The supporters of the virino hypothesis were – and still are – keen to point out that this would explain the slightly different forms these diseases take in different organisms. The prion hypothesis has yet to explain how these different 'strains' develop. The prion supporters argue back that, like the prion, the virino has never been found. What's more, a virino would be destroyed by enzymes or radiation blasts, while the infectious agent for scrapie and the other diseases emerges unharmed from such attacks. Aha, say the virino people, we know of plenty of viruses that can survive enzyme or radiation attack.

And so it goes on. There are all kinds of subtleties in the diseases that neither camp's hypothesis can explain. The evidence, such as it is, does not yet allow us to choose between them. And that is why Prusiner's 1997 Nobel Prize caused such a stink.

After the Nobel Prize announcement, members of the Karolinska Institute, the committee that decides on the recipients, found themselves in the awkward position of having to defend their choice. Lars Edström, for example, acknowledged that there were people who don't believe that a protein can cause the family of diseases that includes kuru, scrapie and CJD. 'But we believe it,' he told the *New York Times*. 'From our point of view, there is no doubt.'

Many scientists immediately expressed outrage that the Nobel Prize committee was invoking faith, not facts. Bruce Chesebro, who led the research effort into the causes of scrapie, kuru, BSE and related diseases at the Rocky Mountain Laboratory, issued a press release outlining his objections. 'No one knows what a "prion" really is,' he pointed out. If the award of the Nobel stopped people looking for a virus, he said, the result could be 'tragic'. Laura Manuelidis made a similar point: she told the *New York Times* of her fears that debate would now be stifled. In the same article, Robert Rohwer, a research director at the Veterans Affairs Medical Center in Baltimore, went further and compared prion science to the cold fusion debacle that had rocked physics in 1989, when two researchers caused an enormous stir by claiming to have created energy-releasing nuclear reactions at room temperature. No one has ever managed to replicate that result, and the physicists in question lost their careers over the issue. In *Science*, Rohwer called Prusiner's prion hypothesis 'the cold fusion of infectious disease'. In other words, it might be radical and appealing, but it is entirely unproven.

A few months after the announcement, Chesebro repeated his appeal for prions to be taken with a pinch of salt. This time it was in a paper in *Science* that outlined the reasons why the Nobel committee's award for the prion hypothesis should not be closing the book on diseases such as scrapie and CJD. 'Clearly, we are in the very early stages of exploration of this subject,' Chesebro said. 'It would be tragic if the recent Nobel Prize award were to lead to complacency regarding the obstacles still remaining. It is not mere detail, but rather the central core of the problem, that remains to be solved.'

Prusiner was used to the criticism by now. The *Science* paper had 'set off a firestorm' and unleashed a 'torrent of criticism', Prusiner wrote in his Nobel Prize autobiography:

> Virologists were generally incredulous and some investigators working on scrapie and CJD were irate … At times the press became involved since the media provided the naysayers with a means to vent their frustration at not being able to find the cherished nucleic acid that they were so sure must exist. Since the press was usually unable to understand the scientific arguments and they are [sic] usually keen to write about any controversy, the personal attacks of the naysayers at times became very vicious.

According to Ralf Petterson, the deputy chair of the Nobel committee, the 'firestorm' of criticism was responsible for worsening the effects of the BSE outbreak in the UK. The BSE crisis led to the slaughter of millions of animals, disastrous export bans on British beef and a political nightmare for the British Government. Nobel committee members made explicit mention of the scientific community's reluctance to accept the prion hypothesis as a factor in the scale of the disaster. The scientists had delayed the political decision about when to take action. 'And by then,' Petterson told Reuters, 'it was too late.'

But Chesebro was right. According to the scientific evidence, both the virino hypothesis and the prion hypothesis were plausible – and neither was proved. That's still true today. Yet Prusiner's lab now receives millions of dollars in grant money, partly as the result of his suggestion of a possible link between the prion hypothesis and dementias such as Alzheimer's disease. Manuelidis is effectively sidelined these days, such is Prusiner's status. In 2007 she expressed her disgust by beginning a paper that outlined the case against Prusiner's ideas with rhetoric borrowed from Oscar Wilde: 'I dislike arguments of any kind,' Wilde once said. 'They are always vulgar, and often convincing.'

Prusiner has certainly been convincing. And many would say that the tools he has used to promote his ideas have been vulgar in the extreme. 'The story, to me, is a hideous replay of the tobacco mosaic virus claim of 1936,' Manuelidis says. She has a point: the prion case is an uncanny mirror of a tale of Nobel Prize-winning anarchy that began before Stanley Prusiner was even a twinkle in his father's eye. The odd thing is, this episode may in fact be the root cause of Prusiner's anarchy.

In 1931, the virologist Wendell Meredith Stanley returned to the United States and settled in New Jersey. He had been working in Munich with the Nobel laureate Heinrich Wieland, but an offer had come in from the Princeton branch of the Rockefeller Institute. Leaving Germany turned out to be a good choice: a few years later, Stanley could boast a Nobel Prize of his own.

At the Rockefeller Institute, Stanley began work on finding ways to purify the tobacco mosaic virus. This had been the first virus to be identified – only a few decades earlier – and had quickly become a workhorse of biologists keen to understand the threat these pathogens posed. In 1935 he published a landmark paper in which he claimed to have turned viruses into crystals a few hundredths of a millimetre long. The breakthrough was headline news – the *New York Times* called it 'Life in the Making' – because it reduced something that multiplied itself, and was thus thought to be alive, to nothing more than chemistry. The virus contained no nucleic acids and blurred the distinction between what was alive and what wasn't. Any thoughts that the virus might be some kind of microbe – something that grew, say, or could replicate itself in the way bacteria did – were pushed out of the picture. The tobacco mosaic virus could be regarded simply as a protein that, astonishingly, could multiply itself in the presence of living cells. Stanley was working in the institute's Division of Plant Pathology, where the virus had always been studied through its effects on living plant matter. Now, it seemed, it could be studied alone, and as nothing more than a protein molecule.

When scientists make extraordinary breakthroughs, other scientists are usually quick to step in and try to replicate whatever has been achieved. Stanley's work was no different: when they read his paper, two British researchers attempted to prepare their own virus crystals according to Stanley's published recipe. Frederick Bawden and Bill Pirie, however, got rather different results.

Stanley had said that his liquid crystal virus contained 20 per cent nitrogen but no phosphorus or carbohydrates. Bawden and Pirie, on the other hand, found 0.5 per cent phosphorus and 2.5 per cent carbohydrate. This was a significant difference in composition: carry out the right chemical routines on Bawden and Pirie's mix, and you will get ribonucleic acid – genetic material, in other words. Stanley's 'protein-only' hypothesis was facing a serious challenge. Bawden and Pirie published their findings in *Nature* in 1936. They questioned whether Stanley really had isolated the tobacco mosaic virus as a crystal. Whatever it was he had isolated, it might not be the infectious agent, they suggested.

In a beautiful piece of quiet anarchy, Stanley then silently backed off from his original claims. He began to assimilate Bawden and Pirie's findings about phosphorus and carbohydrate into his own research. By 1938, he was declaring that tobacco mosaic virus did indeed contain nucleic acid. But somehow he did not change the angle of his research, or retract his original claims, and in 1946 he was awarded the Nobel Prize in Chemistry. The presentation speech describes his work as 'one of the most striking discoveries in modern chemistry and biology' and explicitly states that the prize was for demonstrating that a virus 'actually is a protein'. Thirty years before Stanley Prusiner's Nobel Prize split biologists, Wendell Stanley's Nobel had created a scientific stand-off over the same issue: whether an infectious agent can be just protein. Fortunately, Bawden and Pirie had something with which they could fight the respectability conferred upon Stanley by his Nobel Prize. It's something that Prusiner also has in spades: charisma.

In science, charisma can work as effectively as evidence. In his autobiography, *The Double Helix*, James Watson mentions Bawden and Pirie's gift explicitly. 'No one could match the smooth erudition of Bawden or the assured nihilism of Pirie,' he wrote. And so biologists slowly began to desert Stanley and join Bawden and Pirie's camp. By the early 1950s, the pair were winning out, despite Stanley's Nobel Prize.

While they nullified one Nobel Prize, Bawden and Pirie were sowing the seeds of another. Watson knew them because he carried out X-ray crystallography work on the structure of the tobacco mosaic virus. Watson's contract at Cambridge had been terminated, and he was no longer allowed to work on DNA. However, the tobacco mosaic virus was thought to have a helical structure, and that offered a back door into the secrets of DNA. It was, Watson said, 'the perfect front to mask my continued interest in DNA.'

Bawden and Pirie also inspired Francis Crick. The growing appreciation of the role that nucleic acids played in the tobacco mosaic virus led Crick to his 'sequence hypothesis': that the genetic information about an organism is encoded in these nucleic acids. The sequence of molecules in the acid would dictate what kind of protein would be produced.

In 1953, Crick and Watson discovered exactly how it all fitted together – the famous double helix. Three years after that, they published a paper on the structure of small viruses in *Nature*. They defined a virus as a piece of nucleic acid that carried genetic information. That it wore a protective coat made of inert protein was neither here nor there, they said. Bawden and Pirie had accomplished all they intended: Wendell Stanley was finally defeated.

Stanley did not have to return his Nobel Prize. Evidently, though, he did reflect on the long fight for his credibility and the legacy of his stubbornness. In 1970, he published a paper in which he apologised to some of his colleagues. The paper is called 'The "Undiscovered" Discovery', and it makes electrifying reading – especially when seen in the context of what Stanley Prusiner's efforts have achieved.

In 1944, Wendell Stanley's team at the Rockefeller Institute included two very talented microbiologists. Thomas Francis and Oswald Avery were experimenting on the pneumococcus bacterium, trying to establish how bacteria take up genetic material. Almost a decade before Crick and Watson worked out the structure of DNA, Francis and Avery discovered that nucleic acids could encode and transmit genetic information.

And yet no one – including Stanley – did anything about it. In the last section of Stanley's 1970 paper, the section entitled 'An apology', he admits to having no clue why he didn't see what was before his eyes. 'It is obvious that … I was not impressed with the significance of the 1944 discovery,' he says. 'I have searched my memory and have failed to find any really extenuating circumstances for my failure to recognize the full significance of the discovery …'

To anyone working there at the time, the extenuating circumstances would have been obvious. Wendell Stanley was too busy fighting Bawden and Pirie, trying to safeguard his reputation and legacy in the face of vanishing supportive evidence. Stanley's public apology does not stretch to an admission of scientific anarchy, but 'The "Undiscovered" Discovery' does provide a glimpse of the insights Stanley gained into the way to find success in science. Here is what he had to say about Francis and Avery's discovery:

> Clearly the evidence presented was substantial and the investigators recognized that they had made a significant discovery. Why, therefore, was this great discovery not immediately recognized by the scientific world and why did it not influence the direction of biomedical research? Why did not the discovery that nucleic acid could carry and transmit genetic information receive the recognition that it so richly deserved, for this was a major discovery, one contrary to general thought, and hence one that should have immediately affected scientific thinking in several fields. I am convinced that an unfortunate combination of circumstances was responsible.

One of those circumstances, as we have seen, is that Stanley was locked in a battle with Bawden and Pirie. The other circumstances take us straight to the heart of the matter and encapsulate a philosophy that is central to the strategies Prusiner would later employ. If it is important, new science will overturn the old – but not without a forceful promotional campaign. Stanley continued:

> Perhaps of major importance was the fact that the discovery was quite contrary to the dominant thinking of many years and, hence, required not only a vigorous presentation but also a vigorous and continuing promotion for acceptance. This was not forthcoming. In fact, although the authors made the correct conclusion based on the scientific evidence, they were modest and somewhat cautious in their presentation … no one undertook the task of describing the discovery and arguing its merits and significance before scientific audiences across the nation; hence, several years passed before there was general acceptance.

In other words, Francis and Avery didn't have the required combative tendencies. If you want a Nobel Prize, good science is not enough. You need 'a vigorous and continuing promotion'.

Wendell Stanley died a year after he published 'The "Undiscovered" Discovery'. He was buried in California, where he had been a local hero. He had set up the biochemistry programme at the University of California at Berkeley in 1948. If, following the death of his CJD patient in 1972, Prusiner's reading about viruses hadn't included some of Stanley's writings, it would have been a serious omission. So, was Prusiner influenced by the exhortations of Wendell Stanley that scientific breakthroughs require 'vigorous and continuing promotion' in order to make their mark?

It is hard to know for sure. Prusiner has been unwilling to talk to journalists since 1986, when *Discover* published an article that was highly critical of his methodology. But the circumstances and the contemporary reporting of the birth of Prusiner's prion certainly fit the hypothesis.

Gary Taubes' article in *Discover*, the one that angered Prusiner so deeply, was entitled 'The Game of the Name is Fame. But is it Science?' It opens with a quote from Prusiner in which there is none of the circumspection about the difficulties of scientific progress that Prusiner used in later works. This is a celebration of his branding skills: 'Prion is a terrific word. It's snappy. It's easy to pronounce. People like it. It isn't easy to come up with a good word in biology. One hell of a lot of bad words people introduce get thrown away.'

Taubes interviewed one of the postdoctoral researchers in Prusiner's lab at the time; Paul Bendheim says that Prusiner 'rammed that word down the throats of everybody in that laboratory and in the world'. Bendheim and another of Prusiner's colleagues, Dave Bolton, accused Prusiner of hiring fundraising experts to raise the prion's public profile and help get research money out of private foundations. Taubes quotes Bolton quoting Prusiner: 'If we coin a new term for it, and go out and tell people of the potential link to Alzheimer's, we're going to draw people's attention to this. And we're going to get money.'

Another collaborator, Frank Masiarz, threw in the towel at Prusiner's cavalier attitude: 'By creating the name prion, he clearly wanted to push the entire interpretation in the direction of a protein-only agent. I said there's no point creating a name for something that we don't even know exists yet.' Masiarz resigned as Prusiner's deputy in 1982, just after the publication of the *Science* article that catapulted Prusiner to fame. In that article, Prusiner defines the prions: they are 'proteinaceous infectious particles which are resistant to inactivation by most procedures that modify nucleic acids.'

Given the data, the hypothesis that the scrapie agent is a pro-tein is perfectly reasonable. But here's the rub: Prusiner won't be bound by his own definition. If there's a virus, or a virino involved alongside the protein, he will still be right. Taubes' article ends with a quote from Prusiner that blurs everything: 'I never said it's only an infectious protein,' he says. 'I've never said that in one paper. You'll not find it. I've been very, very careful.'

We think of poets, legislators and journalists as people who care-fully and deliberately use words to advantage. Those who write about science are, in the common mind, just recording the facts. That view is naive, to say the least.

In 1964, *Physics Letters* published a paper by Murray Gell-Mann in which he posited the existence of particles called quarks. Triplets of quarks, he suggested, made up the subatomic particles known as neutrons and protons. There were strong mathemati-cal reasons, to do with patterns and symmetries, for proposing the existence of quarks, but Gell-Mann managed to distance him-self from any accountability for their existence. Maybe, he said repeatedly, they are 'just mathematical', and would never turn up in experiments. Maybe, he said on other occasions, they are 'fictitious'. The particle physicist John Polkinghorne caricatured Gell-Mann's equivocation: 'If quarks are not found, remember I never said they would be; if they are found, remember I thought of them first.'

The strategy paid off: when quarks were found to exist, Gell-Mann won a Nobel Prize. Had they not been found, he would never have lost face because he had consistently blurred the edges of what his quark idea actually was. Prusiner is in the same posi-tion. Whatever the prion turns out to be – protein or something that contains only protein – he will be proved right. His approach may not be scientific, in the usual sense of the word; that's why he lost so many of his colleagues. But it is extremely clever.

And it is deliberate. Carol Reeves, an English professor at Butler University, Indiana, has carried out a study of Prusiner's rhetorical style. His published papers, she says, provide a near-perfect illustration of the power of carefully chosen words. In Reeves' view, Prusiner's stock phrase, 'the triumph of scientific investigation over prejudice', was a clever smokescreen. As we have seen, the science hasn't really decided between the various hypotheses. But Prusiner's rhetoric made it look as though it had. His prose is so dense, and his arguments are constructed in such abstruse and complex ways, that the scientists who were uncomfortable with the prion hypothesis couldn't work out what was wrong with it. One researcher, Sue Priolla, who led the team at the NIH's Rocky Mountain Laboratory, told Reeves that 'most people just read through it and think, "Well, OK, that looks OK," and they move on.'

But Priolla stuck with her misgivings. Here's what she told Reeves:

> I knew there was something wrong with that paper. I kept
> rereading it and looking at the data, and then at how they
> explained their data, and finally, after days of pondering, I
> realized it was the wording in a string of sentences. It was
> easy to overlook because it was so subtle.

Reeves examined the text in question with the careful eye of an English professor. She concluded that Prusiner's whole argument in the 1982 *Science* paper rests on an assumption about things that may or may not exist. She says that the entire paragraph 'is actually theory based upon theory based upon theory, clothed in the armor of scientific syntax, requiring enormous reader energy to untangle'.

The masterstroke, really, was Prusiner's first move: putting a label on something that people were already talking about but hadn't yet given a snappy name. In his 1967 paper, Griffith had said, as understatedly as possible, that it could be a protein: 'The occurrence of a protein agent would not necessarily be embarrassing.' Prusiner, on the other hand, named it: 'In place of such terms as "unconventional virus" or "unusual slow virus-like agent", the term "prion" (pronounced pree-on) is suggested.' He even tells you how to say it out loud, like he's teaching English rather than presenting a scientific argument.

After that, it was plain sailing. Prusiner simply presented the prion in a way that created the impression that it was a well-documented, well-characterised object. Take this statement, for example: 'the properties of the scrapie agent distinguish it from both viroids and viruses and have prompted the introduction of the term "prion"'. Reeves points out that this form of words appears several times in the early 1980s, and is a stroke of rhetorical genius. It talks about 'the properties' of the scrapie agent as if they are clear – and clearly distinct from what a virus might offer. It uses the passive 'have prompted' as if the origins of the word 'prion' lay somewhere other than with Prusiner – and as if everyone has now taken it up as the standard.

In a later article, Prusiner says that some agents must be researched further before they can be 'firmly classified as prions' – as if 'prion' is an established classification. In this same article, a similar boldness comes out as 'All the Prion diseases share many features.'

There is no fighting such persuasive use of language, especially when the definition of a prion remains so ambiguous. 'The attempt to subsume within the single term, prion, both the "protein only" and the "protein with nucleic acid" concepts, has made it difficult to engage in precise dialogue,' virologist Richard Carp wrote in 1985.

Reeves interviewed scientists who admitted that their uptake of the prion terminology resulted from repetition, momentum and confusion; they were bamboozled into it, you might say. Eventually, people stopped fighting. They no longer looked for nucleic acids in the agent. And even if they wanted to, they couldn't get the grant money to do it. Carp says that the prion idea is so firmly established that 'there won't be any new graduates coming out who will even be asking whether there is a nucleic acid in this agent, much less have ideas for how to find it'.

Prusiner knew exactly what he was doing, Reeves says: 'Titles that boldly announce theories as facts, declarative statements about the reality of phenomena whose existence and characterization are under dispute, and speculative statements emphasizing productivity over plausibility are all examples of the clear intent … to manipulate readers' perceptions.'

But given the absence of evidence, he had to. Scientists are highly resistant to new scientific ideas. The celebrated astronomer Tycho Brahe stood against the ideas of Copernicus for his entire life. 'New ideas need the more time for gaining general assent the more really original they are,' said the physician and physicist Hermann Helmholtz. The founder of quantum theory, Max Planck, later lamented over his doctoral dissertation that 'none of my professors … had any understanding for its contents'. Ironically, Helmholtz was among those ignorant professors.

Although it is a natural reaction to shake one's head at all this, in many ways stubborn-headed scientists are doing exactly what they are supposed to. Making progress in science is hard because the onus is on the innovator. The unwritten rule says that new ideas must prove themselves. Scientists can't be reeds blown this way and that by every new fad. Science is a battleground. It is written into the constitution of science that the road to Stockholm will be lined with jeering colleagues.

Interestingly, this has been identified as one of the cultural aspects that has kept science in China and the Far East from leading the world. A philosophical tradition founded on Confucianism, where harmony is the desired state, does not get the job of science done efficiently. Much more powerful is the Western method, derived from the ancient Greek tradition where adversarial debate reigns supreme.

In the scientific fight, both sides should be armed only with experimental evidence. But as we saw in the last chapter, gathering the kind of evidence that gets a radical new idea taken seriously is often incredibly difficult. That is why Prusiner adopted a different tactic: persuasion.

And perhaps we should be glad he did. After this examination of Prusiner's anarchy, it is worth pointing out again that he seems to have been right all along. Though they are still defined as loosely as ever, prions are now widely accepted as the cause of the disease family that includes BSE, CJD, scrapie and kuru. They may also be involved with some other diseases that hit much closer to home.

Today, an estimated 35 million people worldwide have dementia. By 2050, that will have increased to 115 million, according to our best projections. It costs getting on for £400 billion, close to 1 per cent of the world's gross domestic product. If dementia care were to be provided by a single international company, it would be the biggest in the world – bigger than Wal-Mart or Exxon Mobil. No wonder, then, that way back in the early 1980s Prusiner said that there was money in the link between prion diseases and Alzheimer's.

Back then, there was no link to speak of – apart from the fact that both involved degeneration of the brain. Today, it seems there might be. But this is speculative, controversial and emotive ground, and we must tread very carefully indeed when we start talking of a breakthrough.

There are a couple of sketchy links between prion diseases and dementia, and they are to do with what biologists call 'prion proteins'. Prion proteins are not a mystery: we know that our brain cells make them. We still don't know what they are for, however. We do have some clues: knock out the ability of mice to make prion proteins, and they lose some of their sense of smell, for instance. The mice also appear to have slightly reduced responses to stress, and may not generate new neurons at quite the normal rate. Most interestingly of all, the absence of prion proteins seems to make long-term memory more robust. Perhaps that is because mice without prion protein are less susceptible to the accumulation of the sticky, clumped-up molecules in the brain that are a symptom of Alzheimer's disease.

When Alois Alzheimer first presented his new disease – in November 1906, at the 37th Meeting of South-west German Psychiatrists held in Tübingen – he focused on the case of a fifty-one-year-old woman who had suffered with memory loss, disorientation, depression and hallucinations. Her brain had atrophied in places, and there were 'clumps of filaments between the nerve cells'. These are the 'plaques' that now characterise the disease.

In the brains of Alzheimer's patients, chains of acid molecules come together to form long, sticky fibres of a substance known as beta amyloid that clump together in the brain. In 2009, researchers were excited to discover that normal prion proteins seem to interact with the beta amyloid fibres and stop them from forming into plaques. What's more, research has also shown that human brain cells engineered to make more prion proteins also make less of the plaque-forming beta amyloid protein.

The link to Prusiner's efforts is this: the overwhelming majority of researchers into prion disease now work on the assumption that the infection is caused by a prion protein that has taken on the wrong shape. And abnormal prion proteins – ones that have taken on the wrong shape – don't offer any protection against the Alzheimer's plaques.

Misfolding is not uncommon in proteins. Once formed, proteins usually fold themselves spontaneously into a variety of three-dimensional shapes, a little like self-creating origami. What started out as a long, boring string ends up, via a mechanism that remains a mystery to biologists, as an intricate sculpture of bends, waves and curves. This shape is central to the functions that protein will perform.

However, although they are remarkably robust – unfold some proteins, and they will fold themselves back into their proper shape – there are exceptions. The action of heat on egg-white protein is one: it unfolds and doesn't return to its original shape as the egg cools. Instead, it forms a white, misfolded mass of protein that is rather good to eat.

Misfolding is rarely a good thing, though. Emphysema and cystic fibrosis are both the result of proteins that don't fold as they should. And so, researchers suspect, are CJD, kuru and scrapie. If these diseases are caused by misfolded prion protein, it seems plausible that a lack of normal prion protein could also be linked to the formation of Alzheimer's plaques.

It seems reasonable to suggest, then, that Alzheimer's and these prion-borne diseases might be somehow connected. In fact, it seems more than reasonable. But although the link between Alzheimer's and Prusiner's prion diseases might look like an open and shut case, it is not. We don't yet understand how (or whether, for that matter) prion protein really is involved in diseases such as kuru and CJD. Our best guess is that they involve prion proteins that have somehow folded up differently to normal; this mis-folded protein then encourages natural prion proteins to fold in the wrong way, leading to the spread of disease. But we don't know that for sure. It's promising that researchers have made misshapen prions in a test tube and injected them into mice that then went on to develop CJD-like disease. However, it only happened in mice that were genetically engineered to produce huge amounts of prion protein in their brains. Normal mice remained perfectly healthy.

The prion-protein-only idea also fails to explain why different strains of the prion diseases appear from the same infective dose: one mouse might become hyperactive, for instance, while another becomes drowsy. Though misfolded prion protein seems to be involved in whatever is causing these diseases, it also seems to be only part of the answer. So far, the prions have proved infectious only in normal animal models when administered with a soup of 'cofactors': fats and nucleic acids. No one knows whether these are essential ingredients or essential catalysts for a chemical reaction that refolds prion protein in the infected animal's brain. There is certainly still room for the involvement of a nucleic acid element – maybe even for a virus or virino – in the prion diseases.

In the end, none of this will matter to Stanley Prusiner, because he never said what his prion was, and he certainly never said that it was purely protein. Anyway, he has his Nobel Prize for getting us this far. Like Murray Gell-Mann with his quarks, Prusiner created a unifying principle that has proved invaluable in focusing research. But after Gajdusek and Prusiner, it is likely that there will one day be a third Nobel Prize associated with the prion diseases – a prize for actually pinning down the infectious agent that causes them.

No doubt this will entail further anarchy. Perhaps it will be in the invention of a new construct, the new 'prion', to bring researchers together. It is even possible that the anarchy will be in the form of bold and extraordinary experiments, the like of which no normal person would dare to perform. After all, as we are about to discover, people working on matters of life and death don't always stop to ask what the ethics committee has to say.

4

PLAYING WITH FIRE

No pain, no gain

London's Northwick Park hospital looks tired, its dreary façades an uninspiring composition of concrete and metal-framed windows. The hospital had a bright moment when it was opened in 1970 by the Queen, but it quickly became a desolate sight. Perhaps that's why, just a few years after it opened, it was chosen as a location for the iconic horror film *The Omen*.

The hospital's most terrifying moment was yet to come, however. On 13 March 2006, TV camera crews filled the hospital grounds. Images of the building were broadcast to millions of fascinated – and horrified – viewers around the world. What could not be broadcast was the scenes within those walls.

That morning, eight young men who had each been paid £2,330 allowed researchers to inject them with an experimental drug. It was called TGN1412, and had shown promise in fighting multiple sclerosis, some cancers and rheumatoid arthritis. Within a few minutes of receiving the injection, six of the men 'went down like dominoes'. The men tore at their shirts to relieve the fever that had struck them immediately. They vomited, they writhed in agony, they fainted. Their faces started to swell up – press reports referred to them as the 'elephant men'. All six suffered multiple organ failure and were hospitalised for weeks. One had to have toes amputated because of the frostbite-like symptoms induced by the drug.

By the end of the year, the company behind the drug had gone bust, but the tribulations of the six trial patients continued. Mohamed 'Nino' Abdelhady, for instance, was covered in dozens of potentially cancerous lumps that had popped up all over his arms, chest and stomach. The lumps were surgically removed, but the fear of unknown future health problems remained. A year later, others reported memory lapses, stomach problems and severe headaches. David Oakley had been diagnosed with lymphoma.

The UK's Medical Research Council's immediate reaction to the incident was that the risk was worth taking: such clinical trials were essential for the development of new and better treatments. The British Government commissioned a group of scientific experts to report on the lessons that could be learned, but an analysis of this group's report in the *British Medical Journal* is telling. Though things 'could have been done better', their task was supremely difficult, the analysis says: they were charged with 'creating a balance between improving safety without being accused of "stifling innovation"'.

It is an impossible dilemma. Innovation is what scientists do – and it seems they will do it, whatever it takes.

In 2005, ethicists Patricia Keith-Spiegel and Gerald Koocher published a rather enlightening paper. It suggested that a scientific institution's ethical review board, set up to make sure that its scientists comply with a globally agreed ethical standard in their experiments, might be having the opposite effect. They were provoked into researching the subject by conversations Keith-Spiegel had had with scientist colleagues. Though the reports are anonymous, the case studies make salutary reading for anyone who thinks that scientists always follow a rigid moral code.

One investigator collects data by assigning her students research tasks. If something publishable comes up, she asks her ethics board for permission to use data she has already collected for 'nonresearch purposes'. A neat side-step. Another asks her review board for permission to collect data – but doesn't bother waiting for a reply before beginning the process. Yet another omits and distorts elements of his research projects that might cause raised eyebrows on the ethics board.

Then there is the 'prolific publisher' who doesn't bother with the review board because it is a 'rigid and antiscientist authority'. And the investigator who didn't like the changes the board had suggested; he went ahead with the study and declared that he had the ethics review board's support when he submitted it for publication.

It doesn't end there. Keith-Spiegel and Koocher's paper mentions the revenge that scientists take on their ethics board: one investigator found himself in a position to deny promotion to a member of his ethics board who had refused to sanction his research protocol. 'The investigator confided with a degree of smugness that revenge tastes sweeter served up cold,' said Keith-Spiegel and Koocher. Others 'expressed the belief that their deceitful actions were fully justified and necessary in the interest of continuing their contributions to science unfettered'.

In the minds of scientists, science must advance. The truth is that regulation is most likely slowing the pace of progress, according to *New York Times* writer Lawrence K. Altman. 'Who knows how many beneficial drugs are being withheld from the public or remain undiscovered because curious scientists are inhibited from following their scientific instincts?' he asks.

The interesting thing is that scientists have always found a way round such barriers. Altman has written the definitive history of this phenomenon, engagingly entitled *Who Goes First?*. It is packed with tales of anarchic avoidance of the 'proper' way to do research, and proves that, because they are hell-bent on advancing science, scientists can always find one willing research subject: themselves.

The procedure pioneered by Werner Forssmann has probably saved the life of someone you know. Every year, millions of people undergo cardiac catheterisation. It is the standard way to look at how the heart is functioning after a heart attack, chest pain or other indications of a heart problem. A description of the procedure is enough to make you wince: a small cut is made in an artery – often near the groin – and a tube is pushed in, all the way to the heart. It's certainly not the kind of thing you would want to do to yourself.

Forssmann's story began in 1929, when he saw drawings that showed veterinarians accessing a horse's heart via its jugular vein. At the time, the heart was off-limits. Expose it, even touch it, received wisdom said, and a patient would surely die. It was a sensible view: we now know that the touch of a foreign body on the lining of the heart wall can disrupt the cardiac rhythms, causing instant death. But Forssmann was frustrated by the impasse. Little was known of how the heart worked – or of what could go wrong with it. He reasoned that if you could somehow get access to the heart via a vein, then we might learn at least something of its workings. Perhaps we could even use a tube to deliver drugs or fluids directly to the heart.

Forssmann may have had the idea, but he did not have the authority. He was an intern in a small hospital in Eberswalde, 30 miles north-east of Berlin. He approached his boss, the surgeon Richard Schneider, and suggested that the technique might be tried out on dying patients. Schneider said no. Forssmann even volunteered to be the subject. Schneider was having none of it, and forbade any such experiment.

What happened next shows just how anarchic – perhaps 'subversive' is a better word here – a scientist can be. Forssmann knew that the experiment would require sterile surgical equipment that was kept locked in the operating theatre. He tracked down someone who had the key, and proceeded to charm them into submission. Chief nurse Gerda Ditzen didn't stand a chance.

'I started to prowl around Gerda like a sweet-toothed cat around the cream jug.' That was how Forssmann described his first move in the astonishing sequence of events that led to his Nobel Prize. Ditzen was passionate about medicine, and Forssmann exploited her passion: he plied her with textbooks, he talked about surgery with her for hours on end, and eventually, when he thought the time was right, he mentioned the experiment he longed to do. Ditzen eventually agreed to give him access to the necessary equipment – and to her own body as the first experimental subject.

One evening, after the theatre had closed, the pair embarked on their forbidden quest. Forssmann loosely tied Ditzen's arms and legs to the operating table. Then he rubbed her arm, where they'd agreed to make the incision, with iodine to make it sterile. Then he disappeared. Ditzen waited for him to return – somewhat nervously, it can be imagined – but he didn't come back. Access to the equipment was all Forssmann had wanted: he had no intention of putting Ditzen's life at risk. Out of her sight, he made a cut in his own brachial vein and catheterised himself, pushing a length of narrow rubber tubing through the vein, up towards his heart.

The procedure produced a 'burning sensation', he said. Once the tube had reached his shoulder, Forssmann went back to Ditzen and showed her what he had done. She was furious at his deception, but he calmed her down and asked her to help him down the stairs to the X-ray department. Now Forssmann could watch the progress of the tube as he pushed it towards his heart. Nurse Ditzen held up a mirror so he could see what he was doing.

The radiography technician had slipped out of the room, and now he returned with Dr Peter Romeis, one of Forssmann's colleagues. Romeis's first reaction was to attempt to remove the catheter. Forssmann resisted the attempt by kicking Romeis hard in the shins. Eventually, and suffering more pain than the man with the rubber tube in his right auricle, Romeis relented. The catheter had reached the heart; there was nothing to be lost now by taking the X-ray picture as proof of this medical milestone.

The picture was published in Forssmannn's breakthrough paper, which appeared in *Klinische Wochenschrift*. It was accompanied by a barefaced lie about how the research was done. Forssmann's boss, Richard Schneider, had advised him to say that he had tried the technique on cadavers first; in the end, Forssmann supplemented this invention with an imaginary colleague who had started the operation but become too disturbed to continue, leaving Forssmann alone to finish the procedure by himself. It was a fittingly fraudulent finale to the anarchy.

The truth did eventually come out, and Forssmann's colleagues and superiors at the Eberswalde hospital were impressed. They were so impressed, in fact, that they sent him to work with the esteemed German surgeon Ferdinand Sauerbruch.

There is a bitter irony to what happened next. When Sauerbruch found out what Forssmann had done, he dismissed it with a brusque, 'You can't begin surgery like that!' But within a decade Sauerbruch was promulgating some anarchy of his own – and of a distinctly darker hue. He eventually became Hitler's Surgeon General to the Army, a role in which he performed and sanctioned medical research on concentration camp prisoners for the SS. The series of experiments – which included exposing prisoners to mustard gas – was one of the reasons for the drafting of the Nuremberg Code.

After the Second World War, the Allies held war crimes trials in the Bavarian city of Nuremberg. During the proceedings it became clear that members of the Nazi party had carried out horrific experiments on Jews. Japanese scientists had also experimented on prisoners of war, and the Allies had experimented on their own citizens and soldiers. All were roundly and rightly condemned, and science was commanded to put its house in order. And so arose the defining ethical guidelines in medical science: the Nuremberg Code.

According to this code, scientists must not do anything to anyone without the subject's informed consent. The experiment must be purposeful, and likely to yield beneficial results. It should avoid unnecessary suffering or injury, and there should be no chance of permanent damage to the subject. And so on and so forth: these are now the standards we assume to lie behind every medical research programme. What happened during the Second World War is shocking: to us, these guidelines seem like common sense, the kind of rules that any normal human being would unthinkingly follow.

Not that all wartime science was mired in dark deeds. During the First World War, for example, father and son biologists J.S. and J.B.S. Haldane subjected themselves to chlorine gas inhalation in order to find the best respirators for the troops; they unquestionably saved thousands of lives. J.B.S. Haldane went even further: during the Second World War he tortured himself to help British navy divers swim deeper and longer underwater without getting the painful and potentially lethal decompression sickness known as the bends. In his experiments he breathed various mixes of nitrogen and oxygen, and used different rates of decompression. Some threw him into seizures: he spoke of suffering 'extreme terror, in which I may make futile attempts to escape from the steel chamber'. He suffered some lasting injuries: a bout of convulsions permanently compressed his vertebrae, for example. After a decompression experiment he ended up with perforated eardrums – by the time the war was over, Haldane could blow smoke out of his ears.

But these experiments were of vital importance. It was Haldane's research that enabled British commandos to defend and hold Gibraltar in the Second World War, despite Hitler's attempts to take control of this vital fortress, the gateway between North Africa and Europe. Whoever controlled Gibraltar controlled the flow of shipping between the Mediterranean Sea and the Atlantic Ocean.

Werner Forssmann also showed himself to be a brave and honourable researcher, despite his anarchic methods. When he was offered the chance to do medical research on prisoners during the Second World War, he refused. 'To use defenceless patients as guinea pigs was a price I would never be prepared to pay for the realisation of my dreams,' he wrote in his autobiography.

And his pioneering spirit was eventually recognised. Humiliated by Sauerbruch's brush-off, he transferred to urology. Eventually, he joined the army and served on the Russian front. Forssmann ended the war as a prisoner of the Allies, but his breakthrough research paper had a life of its own. While he languished in an American POW camp, two Allied physicians – one French, one American – read about his self-catheterisation and used the idea to develop a technique for diagnosing various cardiac diseases. In 1956 Forssmann joined them in Stockholm, where the three surgeons were jointly presented with the Nobel Prize in Physiology or Medicine.

Scientists are not reckless without reason. They don't go about breaking rules for kicks. But the fact is that the rules are sometimes a hindrance to the creative process of science. And when that's the case, the rules will – make no bones about it – be broken. Why? Because science existed before the rules.

The history of science – particularly of medicine – is crammed with examples of reckless behaviour like Forssmann's. It goes back right to the beginnings of the subject. Isaac Newton's notebooks, for instance, describe a moment of eye-watering recklessness:

> I tooke a bodkine & put it betwixt my eye & [the] bone as near to backside of my eye as I could: & pressing my eye [with the] end of it (soe as to make curvature in my eye) there appeared severall white darke & coloured circles. Which circles were plainest when I continued to rub my eye [with] point of bodkine, but if I held my eye & bodkin still, though I continued to presse my eye [with] it yet circles would grow faint & often disappeare untill I removed [them] by moving my eye or bodkin.
>
> If [the] experiment were done in a light roome so [that] though my eyes were shut some light would get through their lidds there appeared a great broade blewish darke

> circle outmost, & [within] that another light spot whose
> colour was much like [that] in rest of [my] eye. Within
> [which] spot appeared still another blew spot espetially if
> I pressed my eye hard & [with] a small pointed bodkin. &
> outmost appeared a verge of light.

Newton drew an annotated diagram of this strange and potentially dangerous experience (I have removed his references to the
figure). He reports it as if it were the most natural thing in the
world. The chances are that many of his contemporaries were self-
experimenting in a similar manner.

Two centuries later, little had changed, it seems. The nineteenth-
century pioneers of anaesthesia were their own experimental subjects. The American dentist Horace Wells made himself the first
patient to have a tooth extracted under the influence of nitrous
oxide, or laughing gas. Wells had noticed the gas's effects at travelling fairs, where it was demonstrated to great acclaim; he saw that
people who had inhaled the gas felt nothing when they bumped
their shins on returning to their seats.

Other anaesthetics were also self-tested. The Scottish obste-
trician James Young Simpson was a particularly fervent self-
experimenter, sitting with his friends in the late evening, sniffing
compounds from saucers and tumblers spread across his dinner
table. A good breath of a 'ponderous material' known as chloroform knocked Simpson out for the night; when he woke up the
next morning he knew he had found something remarkable and
quickly tried it out on his niece to confirm the effect. A few years
later, Queen Victoria used chloroform to enjoy a painless child-
birth, and the field of anaesthesia had come of age.

At the turn of the twentieth century, American surgeons William Halstead and Richard Hall, the pioneers of opiate anaesthesia, tried everything out on themselves – and ended up as cocaine

and morphine addicts. The German physicians August Bier and August Hildebrandt also enjoyed the benefits of cocaine when they performed experimental spinal anaesthesia on each other. According to reports, they celebrated their success in rendering the lower body impervious to pain with wine, cigars and vigorous kicking of each other's shins.

Then came the Nuremberg Code. But even that didn't put an end to recklessness. That's because the code offers a way out for those wanting to perform dangerous or life-threatening research. Though Part 7 of the code says that no experiment should be conducted where there is an a priori reason to believe that death or disabling injury will occur, it also makes an exception. If the researchers are themselves willing to become subjects, as Forssmann did, the risk becomes acceptable.

An interesting common thread runs between Barry Marshall, whose anarchic science uncovered the cause of stomach ulcers, and Kary Mullis, the LSD-using inventor of gene-copying technology. Both these Nobel laureates describe childhoods characterised by unbridled enthusiasm for playing with the tools of the scientific trade.

Mullis spent his childhood building rockets in his backyard in Columbia, South Carolina. He was, in his own words, 'a little scientist guy'. He and his brother developed a chemical propulsion system that would launch a rocket into the sky – carrying the family's pet frog. They knew it wasn't quite the right thing to do, but that wasn't going to put them off. Six months of covert research got the frog to where the boys felt sure it wanted to go. 'We weren't frightened by it, but we should have been,' Mullis says.

As a child, Marshall was similarly intrepid and wily. By the age of eight he was making electromagnets and buying chemicals from a local pharmacy in Perth, Western Australia, in order to manufacture explosives. He played with electronics and electrical equipment. On one occasion he repaired a fraying power lead on his father's drill, but mixed up the wires – a mistake that caused his father to jump a few feet into the air the next time he used the drill. Marshall's father also suffered when he suggested to his son that it was a bad idea to make lighter-than-air balloons using domestic house gas: in demonstrating the danger to his son, he singed off his own eyebrows.

No doubt Mullis and Marshall's neighbours, halfway across the world from one another, were all thinking the same thing: 'That child is reckless and out of control!' But they hadn't seen anything yet. Whereas Mullis manufactured psychedelic drugs that would help him to solve mind-bending problems, Marshall grew up to be anarchic in a different way. Decades on from the explosives factory, the reckless Marshall was experimenting on himself. That was why, in 1984, Robin Warren screamed down the phone to a reporter from the *New York Star*: 'Barry Marshall has just infected himself and damn near died!'

It was Warren who started Marshall on his path to a Nobel Prize. In 1981 Marshall had to complete a research project as part of his medical training at the Royal Perth Hospital. He felt drawn to gastroenterology, the study of the digestive system, and asked around to see if there was anything interesting going on. He was directed to the basement. There he found Robin Warren and his cigarillos. Warren is now retired from medicine, but Marshall paints him as something of a maverick in his time. They spent many afternoons together in that hospital basement, Marshall says, drinking strong black coffee and smoking, and trying to figure out the meaning of Warren's curiously infected patients.

In the course of his research, which involved taking cell samples from the linings of people's stomachs, Warren had found that many of his patients were infected by strangely curved, almost spiral-shaped bacteria. This caught Warren's attention – the overwhelming majority of bacteria are spherical or straight, and there were only a few reports of spiral bacteria. *Treponema pallidum*, the bacterium that causes syphilis, is one. But since his samples of spiral bacteria appeared to be thriving in the hostile acids of the stomach, Warren thought they might merit further investigation: perhaps they were responsible for specific stomach-related health problems. He didn't have time to study the patients himself. But Marshall was welcome to do it.

Over coffee and cigarillos, Warren explained the intricacies of the stomach to Marshall: how bacteria survived the acid by living beneath a thick layer of mucus, or by secreting urea (an alkali) to create a pH-neutral bubble around themselves. Warren handed Marshall the case notes of twenty-seven patients infected with the new bacterium. The only one with an interesting medical problem was someone Marshall had seen on his rounds: a fifty-year-old woman with undiagnosed abdominal pain. All examinations had failed to find anything wrong with her. The single anomaly in her notes was this infection with a curved bacterium.

Good scientists are like TV detectives: give them a few clues and they will chase down the truth. And, as with the most compelling fictional sleuths, the most successful, world-changing scientists use more than cold, logical deduction. Their appetite for the thrills of the chase leads them to cut corners using their wits, their charm, their unique skills and – just occasionally, when necessity drives them to the edge of reason – their willingness to put themselves in danger.

Marshall started his sleuthing in the hospital's library, following up the leads Warren had given him, but he soon found that he would need to look further afield. First, Marshall put his child-hood dabblings with electronics to good use. He built his own computer – this was 1981, the launch year of the Sinclair ZX81 – and used it to make light work of grant proposals and other admin tasks. This not only freed up his time, but the expertise in comput-ing that he acquired allowed him to get in touch with researchers overseas and access research materials in a way that almost no one else in Australia could at that time. Warren could not have assigned the Case of the Curved Bacteria to a better detective.

By the end of the year, Marshall had traced the spiral bacteria back to their first observation. In 1892, an Italian doctor named Giulio Bizzozero reported to the Turin Academy of Sciences that the gastrointestinal tube contained strange, helical organisms which were visible under the microscope. Science is not always efficient, however, and Bizzozero's discovery, published in Italian, was forgotten and rediscovered several times through the next century.

In the 1940s, a surgeon dealing with stomach ulcers and can-cer at the Harvard Medical School observed that almost half of the stomachs he encountered contained spiral bacteria. Unfortu-nately, in the following decade another surgeon undermined the discovery. Eddie Palmer of the Walter Reed Army Medical Center in Washington DC tried to find the helical bacteria by perform-ing biopsies on more than a thousand stomachs. He saw none, and claimed that they existed only as contaminating infections in autopsied corpses.

In 1967, the Japanese physician Susumo Ito discovered them again, this time while studying the stomach contents of cats while he was working at Cornell Medical School. The cats were not ill, and when Ito performed a biopsy on his own stomach, and found the same spiral bacteria, he assumed them to be harmless – he, after all, was perfectly healthy.

As it turns out, Ito was typically Japanese in carrying an infection by spiral bacteria. Almost all his compatriots had the bacteria in their stomach, which now makes sense of the fact that Japan has the highest rate of gastric cancer in the world. The spiral bacteria are far from harmless.

At first, it was only a hunch. The woman Marshall had examined, the one who had turned up on Warren's list of patients infected with the spiral bacteria, had complained of nausea, stomach pain and headaches. Marshall had found nothing wrong with her; though she had a history of gastric ulcers, an endoscopy showed that there was no ulcer in her stomach now. Nevertheless, he was intrigued by the revelation that this woman did have something strange in her stomach – and symptoms of disease to go with it. Could it be that the spiral bacteria did cause hitherto unnoticed harm?

After a few months of reading and thinking, Marshall came up with a plan. He would identify 100 people who were due to undergo endoscopy, and get the endoscopists to take extra samples of stomach tissue from them. Once he had the samples, he would examine them to see how common spiral bacteria infection really was. Then he would try to grow the bacteria on a Petri dish and see if they could be linked with any diseases. He would also try to establish where the bacteria came from and how the infection took root.

By the time the hospital's ethics board had approved the study, Marshall already had a day job – as a medical registrar in the hospital's haematology department. So it was in his tea and lunch breaks that he hurried off to see the endoscopists, collected the specimens and rushed them across to the microbiology and pathology labs. It took months to gather, but by June 1982 Marshall had all the information he needed.

Not that it all went smoothly. The first step was to persuade the bacteria to grow from the samples. The microbiology lab had been given instructions on how best to grow these gut organisms, knowledge that Marshall had gained from a specialist in chicken diseases at the University of New South Wales. But months passed without any success. Marshall's project might have fallen at this first hurdle, were it not for the intervention of the Easter holiday and a superbug infection.

An antibiotic-resistant strain of Staphylococcus had been found in one of the hospital's wards. The managers, anxious to know whether the hospital needed to invest millions of dollars in new antibiotics, had set up a quarantine for medical staff who had been in the area where the outbreak occurred. Their throats were being regularly swabbed for the new strain to see whether they were carriers. The burden of analysing these extra samples fell on the microbiology lab. Dr Marshall's unimpressive project was now a low priority.

In the months leading up to the outbreak, the results of Marshall's experiments had always proved disappointing for the microbiology staff. There was never anything to see, and the samples were thrown away after forty-eight hours. But now the lab staff were under pressure. And that was how the samples from patient number 37, a middle-aged man with a history of duodenal ulcers, came to be forgotten when they were due for inspection one Saturday morning.

Patient 37 had undergone his endoscopy on the Thursday before Good Friday. As usual, Marshall had rushed the stomach lining biopsies to the microbiology lab as soon as he could, and the hospital's technicians immersed them in the prescribed nutrient solution and put the samples into the tightly controlled temperature and humidity conditions of the incubator. And, thanks to the holiday and the extra workload, there they remained, untouched and uninspected, until the Tuesday after Easter. And five days was plenty of time for the spiral bacteria to grow. At last, Marshall had a culture.

Such tales of serendipity are common in science. As we have seen, so are little indiscretions with data: the process of science is just too messy for these to be avoided. That's why Barry Marshall, months before he got to the anarchy that make him the prime mover in this chapter, had to perform a little anarchy on his data. It was hardly anything, but, given what we have seen so far, interesting. No one now disputes the discovery, but it would have been a little less convincing at the time had he not bent the rules by just a few degrees.

The procedure for analysing the data from the patients was exemplary. Rose Rendell, a statistician, was appointed to oversee the analysis. Marshall, Robin Warren and John Pearman, the head of the microbiology lab, sent their results to the statistician directly, rather than giving each other the chance to influence anything. In June, the endoscopists finally sent Marshall their reports of whether they had seen lesions (ulcers) in the stomach walls of the patients. Marshall forwarded the reports to Rendell. By September, Rendell had put everything together and sent it to him. The initial results were exciting.

Of the twenty-two patients with gastric ulcers, eighteen were infected with the spiral bacteria. And, Marshall was thrilled to find, there was an explanation for the other four. All the patients had filled out a questionnaire about their medical health. According to the answers, the four patients with a gastric ulcer but no spiral bacteria were all taking anti-inflammatory drugs such as ibuprofen. These are known to cause stomach problems, including ulcers. Even more satisfying was the fact that nigh on 100 per cent of the patients with duodenal ulcers – twelve out of thirteen – tested positive for this new bacterium. It seemed extraordinary, but the explanation was almost perfect.

Almost. Experimental data don't always play ball, as we have seen. The one duodenal ulcer that was not associated with a spiral bacterium infection could not have been due to anti-inflammatories. Entry to the duodenum is controlled by a gateway, the pyloric sphincter, and there was no way the drugs could get into the duodenum in high enough concentrations to cause an ulcer.

Marshall was worried that he may have miscounted the number of duodenal ulcers: perhaps the one without spiral bacteria present would turn out to be not a duodenal ulcer, but something else. So he went back to the reports the endoscopists had sent him. His persistence was rewarded. The 'anomalous' patient had undergone previous surgery to remove a large specimen from her stomach, and this larger specimen had tested positive for the spiral bacteria. Now a full 100 per cent of Marshall's duodenal ulcers were associated with the bacteria. He sent the revised data to Rendell.

More recent studies have placed the correlation between duodenal ulcers and spiral bacteria at 92 per cent – so twelve out of thirteen is perfectly acceptable. Marshall, though, was determined to make his case – even if it involved a little 'normal misbehaviour'. As the clinician reporting the presence or absence of ulcers, it was not Marshall's place to make a unilateral addition to the infected group. Warren was the one who should have declared whether bacteria were present or not. That is, no doubt, why Marshall writes in his Nobel autobiography, after mentioning his addition, 'Later, Robin must have rechecked the samples and agreed that bacteria were present.' We can regard this as another example of the way things have to be done to get at the truth – and that line of Marshall's is akin to something that Stanley Prusiner might have said. Warren actually has no memory of the incident.

Unearthing this kind of link allows a scientist to form a hypothesis, but it takes courage to stand up for it. Marshall had evidence to show that duodenal ulcers are tightly associated with the presence of a new, as yet uncharacterised, bacterium. While Agatha Christie's Hercule Poirot might leap from such evidence straight to an accusation, a scientist has to be much more circumspect, and Marshall was forced to spend a couple of years carefully making his case.

He had already made a connection between antibiotics – bacteria killers – and cures for stomach problems. He had even tried it out on a patient: an elderly Russian man with 'intractable abdominal pain'. An endoscopy revealed spiral organisms in the man's stomach, and Marshall put him on tetracycline, an antibiotic. Even with the consent of the patient and his physician, this was ethical quicksand – and Marshall knew it. 'This was the first time I realised that our clinical research project was probably overstepping the bounds of what would normally occur,' he later wrote. 'Taking an unnecessary biopsy was one thing, but using obscure findings from that specimen to justify antibiotic therapy was another. Nevertheless, we pressed on …'

It was time to do a study that *would* stand up to scrutiny. The focus of Marshall's attention was now bismuth, a metal that had been used in Germany for more than two centuries as a cure for stomach problems. Bismuth was also used to treat bacterial infections such as syphilis. And syphilis was already known to be caused by a spiral bacterium. The circumstantial evidence was piling up.

Marshall conducted his study with some colleagues at Fremantle Hospital. They dipped some small discs of filter paper in a bismuth-containing medicine called De-Nol, then put the papers in the centre of a culture of the spiral bacteria. Four days later, the bacteria had died, showing up in a clear ring around the papers.

The bismuth had killed them. Marshall describes that as probably the most exciting moment of his life. 'It all fitted too perfectly to be a coincidence,' he later wrote. 'I think that was the first time it crossed my mind that we might win the Nobel Prize.'

Next, Marshall treated a handful of stomach ulcer patients with a mix of bismuth and metronidazole, another antibiotic. Four were cured. It sounds as though by now there should have been enough evidence to convince anyone. But there wasn't.

Marshall needed to provide satisfactory answers to a set of questions known to medical researchers as Koch's postulates. Outlined by bacteriologist Robert Koch in 1890, they provide four hurdles to be jumped before an infection can be said to be inextricably linked with a disease. Marshall had to show that the bacteria were present in every case of the disease. He had to be able to get them from the patient, and grow them in a laboratory culture. He had to take these bacteria and use them to recreate the disease in an otherwise healthy host. Then, in the final step, he had to take a new bacterial culture from this infected host. If he was right about the bacteria, it shouldn't have been all that hard.

Evolution has equipped its pathogens well. Think of the way cholera spreads: through diarrhoea loaded with bacteria that contaminate water supplies, ensuring that they reach new hosts. Bacteria and viruses have become highly adept at reproducing within an infected host, and causing that host to behave in ways that pass on the infection to new hosts. The new hosts then fall ill and pass the pathogens on again.

It was up to Marshall to do nature's job, under strict laboratory conditions – and he fell at the first hurdle. At the beginning of 1984, Marshall tried to infect piglets with the spiral bacteria. The infection refused to take. Though it was clear to him that bacteria were causing stomach ulcers, his failure to fulfil even the first of Koch's postulates meant that his colleagues would not take him seriously. They considered his ideas 'far-fetched'. Experimental results that Marshall considered dramatic, they called 'subtle'. Yes, they said, perhaps these spiral bacteria are there, but there is no reason to think that they are *doing* anything – a view supported by the fact that many people carry the bacteria and suffer no ill-effects. In a crude study of samples taken from local blood donors, all of whom were healthy with no reported gastric problems, 43 per cent tested positive for the spiral bacteria. It seemed that infection was commonplace. Worse, none of Marshall's patients could say where they might have picked up the infection. Without a source of infection, Marshall had no medical story to tell.

Marshall is fond of quoting these words of historian Daniel Boorstin: 'The greatest obstacle to knowledge is not ignorance; it is the illusion of knowledge.' Marshall's peers, his superiors and almost his entire professional world *knew* why people got ulcers. They were caused by stress, smoking, genetics, alcohol, poor diet, and so on. It did not matter that this list was a vague set of conditions that covered almost everybody and anybody – and thus meant nothing. And when someone with a stomach ulcer fell outside this set, they were referred to a psychiatrist: it was a psychosomatic problem. Marshall had seen this at the outset, in a patient who appeared on Robin Warren's initial list of curious cases of infection with spiral bacteria. Desperate to make a diagnosis, Marshall's supervisor had sent the woman for psychiatric evaluation. The psychiatrist said she was depressed, and sent her home with a prescription for amitriptyline, an antidepressant.

Just how desperate Marshall was at this point is shown by the fact that, years later, he telephoned that woman. He was curious to know whether she had ever been treated for the spiral bacteria, and whether she still had gastric problems (no and no were the answers). He has never, however, been able to bring himself to contact the patient whose treatment drove Marshall to his most reckless act. What happened within the walls of that hospital was, it seems, much too painful.

Towards the middle of 1984, Marshall admitted a young man who had blood oozing from his stomach. There was so much blood loss that he was receiving daily transfusions. No one could find the cause. And that included Marshall: after samples were taken during an endoscopy, Marshall looked for the spiral bacteria, but found none. There was plenty of pus, the smoking gun of an infection, but no bullet.

A few days later, Marshall tested the man's blood. He found antibodies for the spiral bacteria. What to do? Marshall knew, after all his frustrated efforts, that his currency at the hospital was low. No treatments had worked, the bleeding was continuing and the hospital's surgeons had taken over the case. Time, Marshall knew, was short, but not as short as everyone's patience with Barry Marshall's obsession. 'I detected a certain coolness amongst my more senior colleagues,' he says, pointedly.

It's what Marshall doesn't say that is most instructive. He discussed the case with the registrar, and suggested they try an antibiotic, but without any conviction. Here was a man who had reached his limit. His reputation was damaged, and that led him, it seems, to argue only weakly in favour of an antibiotic treatment. Who can blame him? It's not as though he had convincing medical evidence that he was right. In scientific terms, it was still just a hunch, just a hypothesis. The registrar evidently decided against following Marshall's advice – and perhaps ignoring it was the proper thing to do – and scheduled surgical removal of the young man's stomach.

'I was too upset to go back to see him,' Marshall says. He still hasn't.

So far, this is just everyday medical science: messy and difficult. But now events were about to take a more anarchic turn. Like the detective who knows the guilty party but can't prove it, Marshall decided to take matters into his own hands. He would be the experimental subject. He would put himself in the same position as that unfortunate young man that no one could help. He would do to himself what he had done to the piglets. He would drink a cupful of the bacteria and let nature take her course. And he wouldn't tell his seniors a thing until it was over.

It was not an easy decision. Every medical research experiment – even a self-experiment – is supposed to be scrutinised by an ethics committee before it can go ahead. And publication of the results in a journal is conditional on the experiment having been approved by the ethics committee as taking place with the full informed consent of everyone involved. The experiment must be useful or necessary, and not ridiculously dangerous. Marshall, considering his standing at the hospital, thought he stood little chance of gaining approval for his self-experiment. Not that the committee's refusal would have stopped him, Marshall says. But it would have stopped him from publishing the results, and it would most likely have lost him his job and ended his medical career.

Marshall didn't tell his wife either. Though he suspects she would have been supportive of his goals, she would not have approved. 'This was one of those occasions when it would be easier to get forgiveness than permission,' he wrote in his Nobel Prize autobiography. Neither did he discuss it with his immediate colleagues, whose help he required. They suspected what was going through his head, he thinks. The endoscopist who agreed to take samples from Marshall's healthy stomach, for example, must have had a clue. But a complicit silence reigned.

On 12 June 1984, just before noon, Marshall drank the bacteria, which were in a cloudy, brown, bug-nourishing broth. He didn't eat anything else all day. Three days later he felt a strange, bloated sensation in his stomach. On the fifth day, he was vomiting as the sun came up.

The morning sickness continued for three days. Adrienne, his wife, told Marshall that his breath was 'putrid'. Colleagues confirmed this. He was sleeping badly, and felt tired and listless. After ten days, he asked the endoscopist for some more samples of his stomach tissue to be taken. Again, a 'don't ask, don't tell' policy ruled. Examination under a microscope confirmed that the spiral bacteria were flourishing inside his stomach, which was also coated in pus. The bacteria transferred happily to a culture dish, and continued to thrive. Koch's postulates were being met.

Marshall was lucky. Saving him a great deal of trouble – domestic, physical and political – his body dealt with the infection all by itself. An endoscopy on 26 June, two weeks after he had ingested the bacteria, revealed that the infection was gone. His blood serum contained no antibodies for the bacteria. The reason for this remains a mystery, and a stroke of good fortune: Marshall now admits that the antibiotic he had been planning to take to clear himself out would most likely not have done the job. Instead, it would have led to an antibiotic-resistant strain taking hold. This would have been difficult to get rid of, and Marshall could have been plagued by health problems for a long time to come. He is, he admits, a fortunate man.

We have now reached the point in the story where the criminal is behind bars, and the police superintendent shakes his head and tells the detective he was lucky – he got away with his maverick methods this time. The detective takes the dressing-down on the chin, but his internal monologue tells him he was right. He stands by his reckless methods; he knows he'd do it all again if he had to.

As the story draws to a close, everybody agrees that he's the finest damned detective the force has ever known.

That's not a bad description of Barry Marshall. Subsequent research revealed that the spiral bacteria are spread, like cholera, through the oral–faecal route – most often in young children. They might have a few days of vomiting, then the infection settles down. For the rest of their lives – for the most part – there are no symptoms. Around half the world's population is infected with the spiral bacteria.

In the early 1980s, the answer to persistent stomach problems was often surgery. Pharmaceutical companies made millions of dollars by selling drugs to treat the symptoms. Barry Marshall's reckless anarchy meant that by the 1990s the game had begun to change. Although just 10 per cent of people who were treated the old way were cured, the success rate under Marshall's new regime was 70 per cent. In the developed world, at least, stomach surgery is now rare, and family doctors, not hospital doctors, deal with ulcers. In 1994 the US National Institutes of Health declared that the first line of attack for doctors presented with a peptic ulcer should be to identify and eradicate the spiral bacterium known as *Helicobacter pylori*.

Barry Marshall's recklessness is by no means a unique phenomenon. Entire books have been written about scientists who experiment on themselves, *Who Goes First?* being probably the most famous. The stories of those who are reckless with the lives of others make less comfortable reading. In 1900, for instance, a group of US Army researchers agreed that they would all suffer mosquito bites to test whether mosquitoes were responsible for transmitting yellow fever, which had become a problem for the army in Cuba during the Spanish–American War. Shortly afterwards, their leader, Major Walter Reed, left Cuba and headed back to Washington. Though many have tried to defend Reed's departure, the evidence points to an unwillingness to put himself at risk – even though he had agreed with his men that they would all be in it together. In his absence, one of the group, Jesse Lazear, died of yellow fever.

Then there is the shameful US Public Health Service project known as the Tuskegee syphilis study. For forty years, between 1932 and 1972, impoverished black men were the unwitting subjects in a study of syphilis. They were not told they had the disease, and they were not given penicillin when it became available in the 1940s as a standard and successful treatment. They were even prevented from attending other treatment centres.

The men were led to believe that they were already in a treatment programme. Instead they were condemned to, in many cases, a lifetime of illness and abuse. Dangerous and painful spinal taps, of use only to the scientists, were offered as 'special free treatments'. The scientists offered the men free transport to the medical centre, and free meals. They also offered to pay funeral costs if the men allowed their bodies to be autopsied as part of the study. The 399 men involved fathered nineteen children with congenital syphilis, forty wives were infected with the disease, and more than a hundred of the men died, directly or indirectly, from their untreated infection.

An equally shameful but more recent example of scientific recklessness with the lives of others is the case of the disgraced British doctor Andrew Wakefield and his discredited search for a link between autism and the measles, mumps and rubella (MMR) vaccine. In the 1990s Wakefield was given £50,000 by a group looking to establish scientific evidence that the MMR vaccine caused health problems. Not only did he not declare this conflict of interest when he presented his results for publication in *The Lancet*, but he also showed what the chair of the UK General Medical Council's Fitness to Practise panel called 'a callous disregard' for the suffering of children. He took blood samples from the twelve children in his study while they were at his son's birthday party. Like the Tuskegee subjects, the children were also given painful and risky spinal taps – without the necessary approval of an ethics committee and with no prospect of any benefit.

The outcome of this reckless study was a shrill and false claim that the MMR vaccine can cause behavioural problems. The doubt this raised in the public mind about the vaccine has created a dangerous situation. In some regions, not enough parents are inoculating their children to reach 'herd immunity' – the level of immunity at which, though cases might still arise, a disease won't spread. Measles has again emerged as a killer disease.

The Korean cloning pioneer Hwang Woo-suk has admitted falsification and fraud, but he also stands accused of recklessness. He needed human eggs for his experiments, and took what he could get without asking too many questions. But perhaps the recklessness was not entirely his. The Korean Health Ministry paid women thousands of dollars for 'expenses' associated with donating their eggs – a hugely questionable practice, and one that has since been made illegal. Those involved in the eggs' end use are also not allowed to donate, in case there is undue pressure from superiors. Nevertheless, two of Hwang's female colleagues were so keen for the research to proceed that they donated their own eggs. They made their donations under false names, but Hwang is not guilt-free: when he found out about the eggs' provenance, he kept it from the journal that was publishing his results.

There are plenty more examples of questionable practices by well-intentioned researchers. National Institute of Mental Health psychiatrist Jay Giedd, for example, carried out brain scans on his own children for four years before the NIMH ethics committee found out and put a stop to it. Giedd was trying to plot the way brains change through adolescence, and his children seemed the perfect subjects. The NIMH's Institutional Review Board begged to differ: scientific ethics says that scientists should not use dependents, or those in some kind of authority-based relationship, in their experiments.

And what are we to make of the extraordinary story of Henrietta Lacks? This woman's cervical cancer provided a line of stem cells that has been used in medical research ever since her death in 1951. The 'HeLa' cells were the first human cells to be cultured outside the body, and have seeded a host of breakthroughs for medical science. Scientists have grown more than 50 million tonnes of Lacks' cells, and put them to work in research that has yielded more than 60,000 scientific papers. Every day, on average, ten new studies are published that owe a debt to Henrietta Lacks. They have earned scientists five Nobel Prizes. However, the original cells were removed without her informed consent. Though her cells have helped to make pharmaceutical companies many millions of dollars, the disposable parts of Henrietta Lacks lie in an unmarked grave in Clover, Virginia. As Rebecca Skloot observes in her acclaimed account of Lacks's life, the complexities surrounding the ownership of donated genetic material make it difficult to know whether there is any way to avoid what some see as exploitation: scientists can and do make money and careers out of their subjects.

Scientists who experiment on themselves, though, cannot be accused of gain without pain. What is most interesting about the phenomenon is that the gain is a hope that gets the scientist through the pain. David Pritchard, for example, is still in that 'hope' phase: he doesn't yet know whether it will be worth the 'indescribable' feeling of having worms burrow through his skin.

Nor is it yet clear whether the stomach pains, vomiting and diarrhoea he induced in himself and others will pay off in the way he hopes. It's too soon to tell whether the time spent explaining to airport security staff exactly what was in the glass vials he was bringing into the UK from Papua New Guinea was worth it. At least, unlike his colleague Alan Brown, Pritchard hasn't had to sift through his own faeces in search of imported parasites that can be put to work in the lab.

In the early 1980s, Pritchard, who works at the University of Nottingham, was carrying out biology fieldwork in East Asia. He was trying to find evidence to back up anecdotal reports that people infected with hookworm, *Necator americanus*, were less susceptible to allergies. Allergies are 'over-reactions' of the immune system, and Pritchard wondered whether the worms were somehow able to turn off, or at least turn down, the immune response. If that were indeed the case, chronic allergy sufferers might get much-needed relief from the humble hookworm.

It was much later, in 2004, that Pritchard decided that the best way to find out would be to let the creatures feed on him. He put a dozen or so of the pin-sized worm larvae on a sticking plaster, and stuck it to his forearm. The larvae secreted an enzyme which broke down the molecular structure of his skin. Then they burrowed in. The burrowing produces an 'intense itching', Pritchard says. Once in, though, only the imagination is disturbed by their progress. The larvae were carried through his bloodstream. When they reached his lungs, they broke through the microscopically thin walls of the capillaries and colonised the alveolar spaces. From there they climbed upwards, into his trachea, until they reached the pharynx. Unconsciously, Pritchard swallowed the larvae. Safely lodged in his small intestine, the larvae grew into adults.

You are probably squirming at the thought of this parasitic invasion. Evolution has equipped us to take every measure – physical, chemical and emotional – to avoid such things. We have a natural reaction of disgust at the sight or smell of the human faeces that can harbour these parasites. Seeing (or imagining) the parasites themselves gives us an instinctive recoil that can save us from infection. An itch on the skin is similarly cautionary: that urge to touch it is a reaction that can swipe away danger. If none of these work, and the parasites do penetrate our defences, our immune system springs into action, attacking foreign invaders, causing us to cough to expel the alien presence in our throat, or to vomit if it is in our stomach.

But these parasites have evolved too. Once inside the human body, they turn off its alarm system – or at least they turn down its volume. By some as-yet unknown chemical mechanism, hookworm lavae suppress the immune response. And that means they get to stay.

That is why, with only a dozen or so worms inside him, Pritchard felt fine. His body did nothing to expel the adult worms as they passed beyond his stomach to his duodenum. There they attached themselves to the blood-rich wall, and siphoned off a small amount of his blood. Eventually, the males inseminated the females, larvae were produced, and Pritchard pooped them out.

Pritchard told his wife what he was up to. She was nervous about it, but that was then. These days, after dozens of cycles, she knows that a de-worming pill deals with her husband's parasites. When he needs more, he can go to his colleague, Dr Alan Brown, whose body is home to a healthy population of hookworms, gathered – or rather, ingested – during the pair's regular trips to Papua New Guinea (this, Pritchard explains, sidesteps the problem of customs officials wanting to know what is in their sample jars). As Pritchard put it in an unlikely sentence in *The Biochemist*, 'His faecal cultures provide the "snek bilong bel", or snakes that live in your belly, for today's trials.'

It's not the way we normally think of scientists going about their work. But it is, to Pritchard, the best way forward in a field of research that could have a major impact on more than a billion people. The fact that the hookworm larvae can burrow into the airspaces in his lungs without triggering an immune response shows that they can do something we can't: somehow, they damp down the immune response. There is tentative evidence that they desensitise those prone to allergic reactions. In Papua New Guinea, people with a faecal egg count corresponding to around twenty-five worms in their duodenum were less wheezy in the neighbourhood of known allergens.

A couple of dozen worms, it turns out, is a tolerable level of infection that produces no adverse reaction. Pritchard tried doubling it to fifty, and it gave him diarrhoea and vomiting. Doubling it again, it turns out, is a disaster: a hundred hookworms living in your duodenum is an entirely unpleasant experience. It is worth noting (and Pritchard has noted it) that the tolerable level of hookworm infection not only seems to reduce asthma. Blood samples taken from volunteers with this level of infection contain antibodies, which means that a mild case of hookworm infection stimulates the immune system in a way that might be harnessed to produce a vaccine, radically improving billions of lives across the world.

It is still too soon to say whether this line of research will ultimately bear fruit, and further experiments are required. But that's fine: there is nothing about Pritchard's work that is considered to be underhand these days. Once he had experimented on himself, Pritchard was able to persuade his university's ethics committee to approve further experiments involving people recruited from the local area. Some are asthmatics, some have Crohn's disease, and some have multiple sclerosis – a disease caused by a self-destructive attack by the body's immune system. There is no known cure for MS, and treatment options are few and problematic. It may be that an answer will be found in the hookworm larvae that Alan Brown scoops from his poop and hands to David Pritchard, ready for that epic journey from Pritchard's arm to his stomach, via his lungs and throat. Disgusting? Yes. Anarchic? Yes. Worth it? Definitely.

We will leave the last word on the extraordinary recklessness of scientific researchers to the nineteenth-century German scientist Max von Pettenkofer. He was a contemporary of Robert Koch, whose postulates about bacterial infection are still followed by medical researchers. When Koch suggested that cholera was caused by a bacterial infection, Pettenkofer was so contemptuous of the suggestion that he drank – Marshall-style – a spoonful of broth laden with the suspect bacteria.

He experienced stomach pain and diarrhoea that lasted a week, but no full-blown cholera. His luck is enviable. But so, somehow, is his reckless courage. 'Even if I had deceived myself and the experiment endangered my life, I would have looked Death quietly in the eye,' Pettenkofer said, 'for mine would have been no foolish or cowardly suicide; I would have died in the service of science like a soldier on the field of honour.'

Of course, some scientists don't just put a single human at risk. Instead – according to some, at least – they endanger our entire way of life. That is because, to the secret anarchists, society's taboos are there to be broken.

5

SACRILEGE

Breaking taboos is part of the game

To onlookers, it seemed like just another comical academic feud that was coming to a head. It was the morning of 13 September 1973, and on the sixteenth floor of New York's Columbia-Presbyterian hospital a bespectacled, red-faced fertility researcher called Landrum Shettles was hurrying towards his laboratory. Trying to retain his dignity, and intermittently looking over his shoulder, he was half walking, half running. A few seconds later a second figure came bounding down the corridor in pursuit. It was Shettles' boss, Raymond Vande Wiele, head of the obstetrics and gynaecology department. Vande Wiele was, witnesses say, muttering angrily under his breath.

The pair had had many public disagreements and fallings-out, and anyone familiar with the department would mark the scene as the prelude to a long-overdue showdown. It would be a story to be retold around the water cooler for days, a source of gossip and giggles. But the outcome was far from amusing. We know that Vande Wiele caught up with Shettles before he could get to his lab, because what happened next had repercussions that are still, nearly forty years later, being felt across America.

On a bench in Shettles' lab sat an incubator set to body temperature, 98 degrees Fahrenheit. Inside the incubator was a small test tube filled with a dark liquid. This was the world's first attempt at *in vitro* fertilisation – IVF. Shettles had placed a woman's egg, her husband's sperm and a cocktail of blood and water in what he considered to be the perfect conditions for fertilisation to take place. Later that day, the woman, thirty-year-old Doris Del-Zio, was scheduled to have the fertilised egg implanted into her uterus.

It wasn't to be. Once inside the laboratory, Vande Wiele snatched the test tube from the incubator. Shettles had broken every hospital protocol, he said, laying the hospital open to multiple lawsuits. Ironically, though, it was Vande Wiele's action that would lead to the courtroom. The following year, Doris Del-Zio filed a $1.5 million claim against him for emotional distress. She claimed that, by exposing the test tube's contents to room temperature, Vande Wiele had killed her baby. Roe vs Wade, the landmark case that defined abortion law in the United States, was only a year old, and this first, stillborn attempt at IVF became part of the national controversy.

On 12 July 1974, the US Congress imposed a moratorium on the kind of foetal research that Landrum Shettles was doing. The moratorium was technically lifted just over a year later, when Congress mandated that all proposals to research human foetal growth, including human IVF, were to be reviewed by a national Ethics Advisory Board.

In 1979 the EAB finally issued their report on the prospects for human IVF. It was far from a ringing endorsement – hardly a surprise, given the social and political climate in which it had been commissioned. During the review process 13,000 comments had come in from the public, the overwhelming majority of them railing against the technology. Senators and Representatives, too, had issued angry letters arguing that IVF research was immoral and unethical.

The report concluded that 'IVF research is acceptable from an ethical standpoint, including research that does not involve the transfer of the resulting embryo, if and only if the research is designed to establish the safety and efficacy of IVF and knowledge that cannot be obtained by any other means.' But here's the rub. A year earlier, while the EAB committee was deliberating, a Harris and Gallup poll had canvassed public opinion on IVF. Americans now favoured the use of IVF by infertile couples, by a majority of more than two to one. Why this shift? Because science had already rewritten the rules. Louise Brown had been born, and she was a normal, healthy baby.

Her name could hardly be more ordinary, yet it has a lasting resonance in the consciousness of the Western world. And no wonder: Brown was the world's first test-tube baby. That label paints her as some freakish creation from a Frankenstein laboratory. Indeed, some people had said that her 'creators', Robert Edwards and the late Patrick Steptoe, were mad, that they were playing God, and that if they were allowed to continue with their work they would sooner or later create a monster. Their peers simply said that they were destined for failure because they were trying to achieve the impossible. But on 25 July 1978, nine days into the trial of Raymond Vande Wiele, baby Louise was born.

The sense of hope this birth gave to childless couples was almost certainly what made the trial judge decide that Vande Wiele had indeed inflicted significant emotional distress on Doris Del-Zio. Landrum Shettles' science might have fallen short of the ethical standards usually demanded, but it had opened up the genuine possibility of artificially created life. What was once deemed the province of the divine had been brought down to earth. Arthur Caplan, a bioethicist who worked in the same building as Vande Wiele and Shettles (and witnessed the pursuit on the sixteenth floor), puts it beautifully: with Louise Brown, reproduction changed from a mystery to a technology. 'What used to be something that most of our ancestors thought of as determined by God or the gods, all of a sudden got determined by a person or two people,' he said in 2004. Gaining control of our reproduction is, to Caplan, as profound a shift as human beings will ever experience. And it is all thanks to the secret anarchists.

Scientists may be anarchists, but they can still have hearts of gold. The development of IVF was motivated not by scientific arrogance, nor by academic ambition, but by a genuinely positive aspect of humanity: empathy. Robert Edwards and his wife had made friends with a couple who were childless. As he watched this couple play with his own daughters, Edwards felt an extraordinary sense of compassion. Relating the story of IVF in his book *A Matter of Life*, he says, 'The trees bore fruit, the clouds carried rain, and our friends, forever childless, played with our Caroline, our Jennifer.' That was Edwards' spur to change the human experience.

The anarchy lay in Edwards and Steptoe's defiance of the establishment. In July 2010, the journal *Human Reproduction* published an extraordinary article. It was an unprecedented analysis of what happened when Robert Edwards and Patrick Steptoe, the researchers who 'created' Louise Brown, approached the UK's Medical Research Council for funding.

Not surprisingly, plenty of religious groups had objected. But the scientific establishment also balked at the request, and not for the most rational of reasons. According to the article in *Human Reproduction*, one problem was that Edwards and Steptoe were outsiders: 'Steptoe came from a minor northern hospital, while Edwards, although from Cambridge, was neither medically qualified nor yet a professor.' The pair were also accused of seeking too much publicity. The review board concluded that they should slow down and try IVF on non-human primates first: someone ought to prove that it works before putting a human foetus through an experience that could go horrifically wrong. 'I would question the ethics of initiating and maintaining an incipient human life for experimental and scientific purposes,' was how Tony Glenister, an embryologist at the Charing Cross Hospital Medical School, had put it. Edwards and Steptoe, true anarchists, ignored the party line and sought private funding for their research.

The concern that test-tube babies would be born with freakish deformities evaporated when Louise Brown came into the world. She was a healthy baby; she is now a healthy adult. She is not a freak. She is not abnormal in any sense. She is entirely unremarkable, in fact: she is now a postal worker in Bristol and has normal, healthy children of her own. But at the time of her birth Louise Brown was a miracle, and Steptoe and Edwards were heroes or villains, depending on your perspective. These days, IVF babies, and the doctors able to create them, are commonplace. Around four million people have been conceived through IVF, with no adverse effects on their health as a consequence. In 2010 Edwards was awarded a Nobel Prize for his achievement (Steptoe died in 1988).

People still ask questions about Edwards and Steptoe's ethics. Before Louise Brown was born, Edwards had overseen a string of failed efforts to create an IVF baby, each one of which hugely dis-appointed the parents-in-waiting. Some take the fact that Louise Brown's parents weren't explicitly told that his technique had yet to produce a live birth to indicate that Edwards' desire for success overrode his adherence to the medical principle of informed con-sent. And IVF is still not without its critics. It is certainly costly, and some say that failure rates are too high. There are also health risks for the woman whose eggs are harvested for the procedure. Going through IVF leaves many couples emotionally and finan-cially devastated if they fail to conceive. But thanks to Edwards' perseverance, there are also millions of parents whose frustration has been turned to joy. When Edwards won his Nobel Prize, the pages of the Nobel Foundation's website filled with unusually per-sonal messages of congratulations from all around the world.

Perhaps the most remarkable aspect of IVF is that the extraor-dinary circumstances surrounding these births no longer make us view the babies any differently to those born 'naturally'. Scientists trample our taboos into the dust and lead us beyond the borders of the forbidden. And we are glad. Harvard Medical School's John D. Biggers has summed up this sentiment with a frank admission. Given what subsequently unfolded, he is *pleased* that Edwards and Steptoe ignored the official concerns. 'In retrospect, it is fortu-nate that Edwards and Steptoe pressed on,' he said in an editorial accompanying that *Human Reproduction* paper. 'Although the grant was rejected, Edwards and Steptoe's visions and persistence have benefited an enormous number of infertile people, both male and female.'

It is a rarely acknowledged truth that, among all society's leading figures – political, intellectual, social and religious – scientists are the ones who really can take us into the promised land. Although religious groups are often seen as sources of moral and ethical guidance, they actually follow science – they do not lead it. And nowhere is this more obvious than in the sacred territory of life and death.

It was, for instance, the philosopher Aristotle – not the author of some sacred text – who most of the world's major religions looked to in their initial forays into reproductive ethics. In his view, conception took a few days, during which the semen 'set' the menstrual blood, just as rennet sets milk to make cheese. The foetus then went through a succession of souls: first 'nutritive', then 'sensitive', then 'rational'. Truly human life, according to Aristotle, was present in the womb only once distinct organs had formed and the foetus became capable of movement. This 'quickening' occurred at forty days for males, and ninety days for females.

Aristotle had little empirical evidence on which to base these arguments. He based his dates on when women said they felt foetal movements, and on examinations of miscarried or aborted foetuses. But it was the best information available to him, and religious groups made full use of it in framing their own dictates on the ethics of embryology.

The Babylonian Talmud, compiled around AD 500, says that it takes forty days for an embryo to form in the mother's womb. Until then it is 'merely water'; after then, it is 'like the thigh of its mother' – and thus imbued with only a limited humanity. Aristotle helped to set the first Christian time limit on abortion. Pope Gregory XIII's mandate was informed by St Augustine's discussion of the issue; the central questions in abortion for Augustine were 'Does the foetus have a soul?' and 'Is the foetus formed or unformed?'

> If the embryo is still unformed, but yet in some way ensouled while unformed ... the law does not provide that

> the act pertains to homicide, because still there cannot be
> said to be a live soul in a body that lacks sensation, if it is in
> flesh not yet formed and thus not yet endowed with senses.

When that ensoulment happened was simply a matter of opinion. In 1584 the Pope decided to set it at forty days – at, in other words, Aristotle's 'quickening'. Science had, subtly, taken control of the religious viewpoint.

That limit remained in place for 300 years, but science's influence only got stronger. It was a subsequent scientific development, the invention of the microscope, that led to the next change in the Catholic Church's position.

Armed with the microscope, scientists could follow the development of the foetus more closely, and they saw that Aristotle's forty- or ninety-day distinction, as well as his ideas on organ development and foetal movement, were entirely baseless. In 1827, scientists demonstrated the existence of the ovum and began to unravel the process of conception. It quickly became obvious that the process of development from conception to birth was continuous, so to save the soul of the unborn child, abortion simply had to be forbidden. By 1869, Pope Pius IX had declared abortion at any stage of pregnancy to be a mortal sin. By 1917, the Catholic standpoint was that anyone associated with the process of abortion – even for therapeutic purposes, where the mother's life is at risk – faced excommunication.

Science has made the creation of life no less thorny an issue. In 1897, the Vatican had to issue a Papal Bull banning Catholics from attempting artificial insemination. Why? Because that year, Walter Heape of Cambridge University reported that the process had become routinely successful in dogs, horses, foxes and rabbits. In 1968, the march of reproductive technology, particularly the contraceptive pill, raised 'new questions which the Church could not ignore', Pope Paul VI declared. As a result, he mandated that there is an 'inseparable connection, willed by God, and unable to be broken by man on his own initiative, between the two meanings of the conjugal act: the unitive meaning and the procreative meaning.' The emerging reproductive technologies were not for use by Catholics.

It was an understandable edict. The problem the Church has is that human beings are disturbingly creative and pragmatic. As a species, we innovate, then happily assimilate the innovation into our normal lives – especially if it solves a pressing problem. And perhaps there is no area of human endeavour more pressing, and less likely to be held back by anything other than practical difficulties, than reproductive science.

Actually, IVF caught the Catholic Church somewhat unawares. At the time of Louise Brown's birth, the Church had said nothing specific about the technology. It has since banned its members from using IVF, declaring it 'morally unacceptable'. But the Pope might just as well have saved his breath. The Catholics – whether they be doctors, scientists or infertile couples – were never listening.

After a further Vatican pronouncement on the immorality of IVF in 1987, several European Catholic hospitals announced that they would defy the edict and continue with IVF. It was, one hospital said, 'an infinitely precious human service'. Margaret Brooks, an Australian Catholic and the first woman to have a child born from a frozen embryo, told the *New York Times* that no one paid such pronouncements any attention. In 1985 a poll found that 68 per cent of American Catholics approved of artificial contraception. And in 2005 a study by the Genetics and Public Policy Center at Johns Hopkins University suggested that Catholics also strongly support IVF.

You can blame the secret anarchists for that. Every proposed innovation in this area has initially been rejected as foolhardy at best, or terrifyingly immoral at worst. But scientists have pressed on regardless – that's what they do. And when science provides a way to satisfy a biological drive, externally imposed moral principles cannot stand in its way. Cardinal Renato Martino, head of the Vatican's Pontifical Council for Justice and Peace, was asking for trouble when, in the face of IVF success, he repeated the papal declaration that 'no one can be the arbiter of life except God himself'. Scientists are no respecters of popes or cardinals.

Craig Venter provides a good example of that. There would be no point in telling too much of Venter's story here because he has done it so thrillingly in his autobiography, *A Life Decoded*. However, after all the historical examples, it would be looking a gift horse in the mouth to ignore this most anarchic – in the best possible sense – of today's scientists. Especially when he makes no secret of his anarchy.

Venter describes himself as a natural risk-taker, 'rebellious and disobedient', endlessly curious, with an 'insatiable urge to build things'. As a child he roamed free through his neighbourhood, built forts and soapbox racers, used lighter fuel to set fire to his plastic battleships and bought firecrackers to make them explode. His devil-may-care attitude survived into adulthood intact: the many exploits of his navy career include lying and cheating his way out of an unfair court martial. As a student he organised sit-ins and demonstrations. Venter happily admits to trying various drugs; nothing was off limits. Perhaps that is why he took to heart the advice of his scientific mentor, the biochemist Nathan Kaplan, who told him never to talk himself out of doing a 'crazy' experiment.

The encouragement, which came early in his career, stood Venter in good stead. It was this, after all, that persuaded him to try a gene-sequencing technique that everybody said would prove useless. As it turned out, the 'dogma of the day', as Venter described it, was wrong. His chosen technique, called expressed sequence tags, turned out to be a 'very big winner' and the beginning of the race towards sequencing the entire human genome. The story of the eventual publication of the human genome is a long and complex one, and is well described in Venter's book and in James Shreeve's *The Genome War*. But there are a couple of moments worth highlighting.

The first was when Venter gave away 38 million dollars' worth of intellectual property. It happened as soon as he got the chance: in 1997 he managed to separate his company from Human Genome Sciences, who had provided much of his research funding in exchange for co-ownership of the gene data.

Scientists are often forced to go into business – by their universities or other institutions keen to capitalise on their investment. By and large, it is not a comfortable state of affairs: though having enough money to buy freedom for their research is important to them, accumulating wealth does not interest most successful scientists. As English doctor Thomas Browne said in the seventeenth century, 'No one should approach the temple of science with the soul of a money changer.' Einstein expressed the scientist's view with a typical simplicity: 'Science is a wonderful thing if one does not have to earn one's living at it.'

Venter, despite all that his critics had said about him, evidently felt the same. He wanted the freedom to work (and salaries for his staff), but beyond that it was all about discovery. As soon as his company had divorced itself from the moneylenders, and all of the genetic data that his work had revealed was his, he made the largest single deposit in GenBank, the publicly accessible database of genetic information. Many biologists working in the public sector rejoiced.

The second illuminating moment of Venter's effort to sequence the genome demonstrates his desperate urge just to be able to get on with it – and let the consequences take care of themselves. The requirement for informed consent meant that Venter was faced with a six-month wait to start sequencing if he used other people's DNA. Impatient and unencumbered by a need to follow the rules laid down by his company's scientific advisory board – not to the letter, at least – Venter and his colleague Hamilton Smith kick-started the project with their own DNA. They kept that as quiet as possible for as long as possible, because they knew it was far from ideal. 'We could expect political attacks from our detractors if it became known that we had used our own DNA,' Venter says.

Eventually, though, the secret leaked out. Venter's company, Celera Genomics, said as little as they could about it. Although its board of scientific advisers expressed some disappointment that Venter had not played exactly by their rules, there was no expression of surprise. Even Venter's competitors and detractors were largely unwilling to score points; despite Venter's fears, their reaction was more catty sniping than political attack. On being told that the decoded human genome was 60 per cent Craig Venter's, DNA pioneer James Watson told the *New York Times*, 'That doesn't surprise me; sounds like Craig.' When the same article's writer suggested that Venter's body should be preserved along with his genome, Stephen Warren, editor of the *American Journal of Human Genetics*, said, 'That would be his wish, no doubt, to be prominently displayed in the Smithsonian.'

What is interesting is that no one expressed shock or outrage. Scientists, particularly those who work with Venter, have come to expect such anarchy. After all, this is a man who can quite justifiably be accused of playing God.

'This is the first self-replicating species we've had on the planet whose parent is a computer.' Those were the words, spoken by Venter at a press conference on 20 May 2010, that sent the President of the United States of America into a barely concealed state of panic. Venter's science raised 'genuine concerns', Barack Obama said. Within a few hours of Venter's announcement, the President had initiated a study to look into its implications.

What Venter had done was certainly extraordinary. If the announcement had come on a videotape from Osama bin Laden, it would have triggered widespread alarm. Venter's research team had created the world's first self-replicating synthetic genome. The design came from a computer, and the genome was created from chemicals stored in glass bottles. Once assembled and placed inside a scooped-out bacterial cell, the instructions of this genome were carried out to the letter. The cell split and reproduced itself – the split-off cell contained another copy of the artificial genome – and carried on its life as if it had been on Earth for millions of years, rather than just a few days.

Venter's aim is not to strike terror into the hearts of humanity. He hopes that this synthetic cell – named Synthia – will be the first of many diverse artificially produced bacteria that become the new chemical factories. The DNA instructions inside many naturally occurring bacteria cause them to produce a range of chemicals. Biochemists have already learned how to turn this ability to our advantage by engineering bacteria that produce human insulin, for example. Venter wants to develop bacteria that can produce synthetic petrol, or process the excess carbon dioxide that is contributing to global warming. These bacteria are meant to save the world.

Not that Venter is working at quite that level yet. The synthetic genome he created was just a copy of the genome of an existing bacterium. The natural *Mycoplasma mycoides* bacterium that lives in cows and goats is almost identical to the synthetic *Mycoplasma mycoides* that Venter's team made. The only difference was that Venter's team had inserted a few 'watermarks' into the genome. Using the chemicals of DNA, they wrote a web address and a few famous quotations in the space after the naturally occurring genetic information. There was a quote from the late Caltech physicist Richard Feynman: 'What I cannot build I cannot understand.' James Joyce made it in there too: 'To live, to err, to fall, to triumph, to recreate life out of life.' The web address was for the use of anyone who managed to decode this information. One can only assume that there is a prize for the first to show up having cracked the code.

It's a playful touch, which is perhaps why many of Venter's competitors have accused him of hype and gimmickry. Yes, Venter had built a genome from chemicals, they said, but he had to insert it into an existing bacterium. 'He has not created life, only mimicked it,' Nobel laureate David Baltimore told the *New York Times*. The work is 'an important advance in our ability to re-engineer organisms', said biomedical engineer Jim Collins in *Nature*, but 'it does not represent the making of new life from scratch'.

The philosophers and ethicists were more alarmed. According to philosopher Mark Bedau, it was 'a defining moment in the history of biology and biotechnology'. Bioethicist Arthur Caplan, the same man who saw Landrum Shettles being chased down a hospital corridor by his boss, said that there is 'great need for more oversight of this hugely powerful technology'. In the wrong hands, Caplan pointed out, it 'could pose serious risks to our health and environment'.

There were no end of comments, in fact: Venter has seemingly limitless power to provoke. Most impressive, though, was the response from the White House. In a letter dated 20 May 2010 – the same day as Venter's press conference and publication in

Science – the President requested the Presidential Commission for the Study of Bioethical Issues to undertake a study of 'the implications of this scientific milestone, as well as other advances that may lie ahead in this field of research'.

President Obama specifically wanted faith communities to be asked for their perspective on Venter's work. The request is unsurprising, given the nature of the work. But all the faith communities will be able to do in the face of Venter's work is to accept it as a defeat or to splutter and frown, for he has pulled the rug from beneath their feet. As Caplan wrote in *Scientific American*, to many scientists, theologians and philosophers, life is 'sacred, special, ineffable and beyond human understanding'. But Venter, Caplan suggests, has now shown that life isn't that way at all. 'What seemed to be an intractable puzzle, with significant religious overtones, has been solved.'

It will only get more difficult for any religion to compete with what science has to offer. Quoted in the journal *Nature*, Georgetown University's Kevin FitzGerald, a Jesuit priest with PhDs in molecular genetics and bioethics, remarks that we are now only at the beginning of the problems for religious interpretation and guidance. When everything we work with to create life has been reduced to molecular formulae, the current issues surrounding stem cell research will seem mundane. 'The stuff that's coming down the pipe will make this look like child's play,' FitzGerald says.

Muslims, for instance, are permitted to use IVF if they are married, but there can be no donor insemination: the sperm and egg must both come from the couple themselves. Donor insemination is regarded as adultery by the wife. That seems pretty straightforward, a nice simple solution, but it is not future-proof. What if the husband is infertile, but sperm could be derived from embryonic stem cells taken from a donor? That would not involve accepting sperm from anyone – would that be acceptable? And what if their baby could be cultured in an endometrium derived from, as in Aldous Huxley's dystopian vision of the future set forth in *Brave New World*, a sow's peritoneum? Where is the line to be drawn?

This is not an idle question: such innovations are coming. A human being self-assembles (in the right conditions) from just two cells: a sperm and an egg. And, thanks to scientists such as Karim Nayernia, we are already learning how to make those.

Nayernia was raised and trained in the region of Shiraz in the south-west of Iran. This is where the oldest known sample of wine, sealed in clay jars more than 7,000 years old, was found. Shiraz is famous the world over as the source of the Shiraz grape: the fruit used to make, say, a Californian Shiraz owe a debt to – and still have some distant genetic link with – the original Iranian grapes that ripened in the Shiraz sun thousands of years ago. Nayernia would argue that it is the same with reproductive technologies.

Just as the Shiraz grape has been exported and combined with technological developments to meet a worldwide demand, Nayernia has taken his expertise in the basics of natural human reproduction, exported it and combined it with some of the greatest innovations in biomedical technology. The result has been an adaptation that will also meet a global demand: Nayernia has made artificial sperm.

The research first came to the attention of the world's press on 13 July 2006. That was the day Nayernia announced that he had managed to produce live baby mice using sperm grown from embryonic stem cells. Stem cells that have been taken from a newly formed embryo are able to turn into any of the 200 or so types of cell found in the body. Because such stem cells can reproduce themselves without making this transition, though, researchers have identified them as of enormous potential as utility cells: they proliferate happily, and can be anything you want them to be.

To make his breakthrough, Nayernia had harvested stem

cells from a mouse embryo, allowed them to develop a little, then removed the few that had transformed themselves into the progenitors of sperm cells. Once these had developed into fully fledged sperm cells, they were injected into mouse eggs.

Success depended on playing a numbers game. The researchers injected a total of 210 eggs with artificially grown sperm. Only sixty-five of those were properly fertilised and began dividing. Each one of these was implanted into the uterus of a mouse, but the process yielded only seven live births. Of those seven baby mice, six survived to adulthood.

Less than a year after the announcement of his creation of functioning sperm cells from embryonic stem cells, Nayernia announced the next step forward: sperm from adult human stem cells. Turning the embryonic stem cell breakthrough into a viable and useful technology would require making very large numbers of cloned embryos of infertile men – which for some people would raise an ethical dilemma. But use adult stem cells, such as those found in bone marrow, and you wouldn't have to be cloned to get yourself some working sperm.

Adult stem cells are not as versatile as embryonic stem cells: they cannot develop into absolutely any kind of cell. But take them from the right place in the body and you can generally get what you want. Nayernia took bone marrow stem cells from adult men and nurtured them into spermatogonia – cells that will, given the right conditions, develop into sperm. These cells have to go through a series of developments before they can become sperm cells, including three bouts of division, known as meiosis. Nayernia has done this with mice, and has made preliminary – and contested – claims to have achieved it with human sperm. But regardless of whether he has achieved this, his work seems to show that there is no fundamental barrier to culturing sperm that work. It is only a matter of time and money. It doesn't sound as though the creation of a new life is going to remain a miracle for long.

The sperm are only one side of the equation, of course. Whereas sperm can just be grown and injected into an egg, making the egg itself is a more complex proposition. Many months before a girl is born, primordial follicles have formed inside her. These are containers for the undeveloped version of the oocyte, the cell that can eventually mature into an egg. The primordial follicles are in a kind of suspended animation: their development is halted, but it will resume at any time between the onset of puberty and the beginning of the menopause. The steady stream of developing follicles is what gives a woman her fertility: each one of them has the potential to grow an egg to 200 times its original size in just a few months, and then burst open, releasing the egg into the Fallopian tube, beginning the process that can end with a new human life.

In 2003 a research team at the University of Pennsylvania announced that they had grown something resembling eggs in a Petri dish using embryonic stem cells harvested from mice. The eggs showed some signs of attempting cell division to prepare for maturity, but things went no further. Others have since achieved the same results, but that, really, is as far as this technique has gone. No one has yet managed to get a fertilisible egg to develop from stem cells. The problem may have to do with the environment in which the eggs develop. In natural conditions the medium in which they grow is a recipe honed by evolution, and both the ingredients and the texture seem to matter. Like industrial food scientists trying to recreate a competitor's market-leading recipe, fertility researchers are performing experiments to see what the essential ingredients might be.

For instance, Alan Trounson, a stem cell researcher at Monash University in Melbourne, has grown cells from testes in a liquid broth, then used that broth to aid the growth of mouse embryonic stem cells. The growth of testes cells certainly seemed to release useful growth factors into the liquid: the stem cells steeped in it developed into something akin to egg-carrying follicles. Renee Reijo Pera, a researcher at the University of California at San Francisco, added bone proteins to her stem cell mix and found that this boosted the number of stem cells that became proto-eggs. Teresa Woodruff of Northwestern University, Illinois, has pushed the texture side of things: she made capsules of gel from a chemical found in seaweed and injected natural harvested mouse follicles into them. The gel provided a perfect supporting medium: when Woodruff stimulates ovulation with a hormone injection, fertile eggs are released from the follicles.

Woodruff has managed to fertilise these eggs, and produce live young. In 2009 her team managed to incubate human follicles to the point where they looked like mature eggs that produced all the standard hormones, such as oestrogen and progesterone. The law still forbids the fertilisation of human eggs for research purposes, but Woodruff is amazed that progress has been so quick that they are now bumping up against this barrier. Ten years ago, she told *Wired* magazine, it looked as though where they are now was more than fifty years away. The procedures involved have 'turned out to be much simpler than we ever dreamed', Woodruff says.

There is still the spectre – as with IVF – of producing babies with abnormalities. The one serious difficulty in creating a new life from scratch lies in a process called imprinting. The DNA contained in every cell in the body is, essentially, one long molecule. Within that molecule are the genes that carry the instructions for making proteins and other molecules essential to the processes of life. In mammals, those genes are 'imprinted' with a chemical tag that switches certain sets of genes on or off, and it is this imprinting that, when a male and female's DNA is combined in an embryo, determines what qualities it will have. So, when creating sperm and eggs from stem cells, imprinting the DNA with the right tags is crucial. Any existing chemical tags have to be removed and the correct new tags have to be put in their place. Get this wrong, and foetal abnormalities are inevitable. At the moment, we have yet to get a proper handle on the imprinting process: what chemical environment creates the right tags in the cultured cells, for example, remains a mystery. But if the history of assisted reproduction teaches us anything, it is that there is no reason why we should not be able to solve that mystery.

The artificial womb is coming too: financial issues aside, it is now seen simply as an engineering problem. Hung-Ching Liu, a researcher at Cornell University's Center for Reproductive Medicine and Infertility, is building a womb from a few cells taken from the endometria, the womb linings, of women visiting the clinic. The clinic's considerable IVF success rate, Liu figured, might be improved by giving the embryo a few home comforts – even if they were contained in a Petri dish.

In Liu's first attempts, the endometrial tissue was too thin for the embryo to implant onto. It kept breaking through and hitting glass, like a tree root hitting bedrock, and contact with a hard surface is enough to stunt the development of any embryo. Then Liu borrowed an idea from skin grafting: she grew the tissue on a biodegradable scaffold woven from collagen and chondroitin, a major component of the body's cartilaginous tissue. The scaffold was bowl-shaped, but gradually disappeared over time, leaving a bowl of endometrial tissue. Liu then took embryos left over from IVF treatments in the clinic, placed them in the layers of cultivated tissue, and stood back to watch. For ten days they grew and developed.

Having tasted success, Liu removed and destroyed the embryos: according to the US Federal Government's rulebook, two weeks is the maximum time for which you can grow human embryos in a US laboratory. She remained excited by what she had seen, though, and decided to take the research forward using animal embryos instead – mouse embryos, to be precise. This too was successful. But it was also disturbing. Liu made an artificial mouse uterus, just as she had made an artificial human uterus. She placed mouse embryos in its tissues. The embryos implanted happily, and their cells began to divide. The growing embryos sprouted blood vessels. They went almost to full term. And every single one had significant deformities.

Gestation is an incredibly complex process that requires an ever-changing chemical environment. As every researcher trying to build an artificial uterus has found, each stage has particular requirements, and the whole process places enormous demands on our ingenuity. But it is still just an engineering problem.

Ectogenesis, growing a baby outside the womb, will begin the 'third era of human reproduction', according to Stellan Welin, Professor of Biotechnology, Culture and Society at Linköping University, Sweden. The first era was 'normal' conception and pregnancy; the way modern humans have been reproducing for the first 200,000 years of their existence on this planet. The second era began with the birth of Louise Brown in 1978. Here, the foetus begins its life outside a woman and is then implanted into her body. In the third era, which may begin in our lifetimes, the entire gestation of the foetus can take place outside the woman's body. According to fertility researcher Roger Gosden, who worked with Robert Edwards at Cambridge University during the pioneering days of IVF, this will be the boldest evolutionary step for a hundred million years – since embryonic membranes were first adapted to make a placenta and sweat glands were modified to produce breast milk. And it *will* happen, Gosden says. 'The chance at last to understand the most tender period of existence and, even more importantly, to heal diseases and to help with the creation of life will surely prove irresistible.' Arthur Caplan agrees: it might take sixty years, he suggests, but it is 'inevitable'.

The fact that we can discuss all these technologies, that we understand the issues involved and have reduced them to engineering problems, suggests that it will indeed be inevitable. Our command over the processes of life will raise innumerable issues, but the important point to note is that it will happen – and, for all the hand-wringing and soul-searching we might do beforehand, it will not be the end of the world.

When the Austrian physicist Lise Meitner was a child, her grandmother told her that the sky would fall down if she did her embroidery on the Sabbath. In a move that cemented her commitment to experimental results as the best source of reliable information, Meitner decided to test the notion for herself. One Sabbath she tentatively stuck the tip of a needle into her embroidery. Nothing happened. Then she made a couple of stitches. Still nothing. For the rest of her life, Lise Meitner happily enjoyed her favourite hobby seven days a week.

The sky did not fall on our heads as we learned to perform IVF. In 1977, the economist and social analyst Jeremy Rifkin issued an apocalyptic warning about the technology that seems a little ridiculous today: 'What are the psychological implications of growing up as a specimen, sheltered not by a warm womb but by steel and glass, belonging to no one but the lab technician who joined together sperm and egg?' Looking at Louise Brown, born in the following year, the answer is clear: there are no psychological implications whatsoever. Rifkin has since suggested that children nurtured in an artificial womb might become 'violent, sociopathic or withdrawn'. That is a pointless, ignorant and potentially harmful speculation. The human experience changed with IVF, yes. But the sky did not fall down. Neither will it fall down as we pursue other 'miracle' treatments. What will happen is that the human experience will change again, as it did when Louise Brown was born. And, as with IVF, it will be a change for the better.

Much of the hand-wringing over reproductive technology makes reference to Aldous Huxley's *Brave New World*. The book opens in a baby factory, the 'Central London Hatchery', with a frightening scene where we encounter a society that has industrialised the process of human reproduction. Huxley was well placed to write such science-inspired fiction: his elder brother Julian was a widely respected biologist, and the genetics pioneer J.B.S. Haldane was a friend of the family. No doubt the story was fuelled by dinner party conversations of what *could* be done with human reproduction if researchers had carte blanche and the will to implement such a scheme. Huxley wrote chillingly of a crimson darkness 'like the darkness of closed eyes on a summer's afternoon' in which foetuses are grown on sow's peritoneum, and 'gorged with blood-surrogate and hormones'.

The legacy of Huxley's *Brave New World* has been an irrational fear of, and disgust for, reproductive technology. But what those eager to draw comparisons with Huxley's vision and the real-world situation with reproductive technology invariably overlook is that in 1946, fifteen years after the book's first publication, Huxley declared that he would write it differently, given the chance. In a foreword to a new edition of *Brave New World*, he said that he would like to incorporate the idea that 'science and technology would be used as though, like the Sabbath, they had been made for man, not ... as though man were to be adapted and enslaved to them'.

It is clear that, as we always have in the past, we will continue to take up the technologies that the anarchy of science offers us: as they mature, they will become an everyday choice. The advantage of artificial sperm and eggs, when the natural ones are not up to the job, is easy to see. The advantage of an artificial womb seems less clear-cut, but in certain circumstances it may prove safer than a mother's womb. As far back as 1971, Edward Grossman, a lawyer working for the US House of Representatives, pointed out that an efficient artificial womb, 'far from increasing the incidence of birth defects, would reduce them by keeping the foetus in an absolutely safe and regular environment; safe, for example, from infection by German measles or drugs taken by the mother'. For those who have no working womb, meanwhile, it will be a godsend. 'I find ectogenesis in many ways repugnant,' Stellan Welin wrote in *Science and Engineering Ethics* in 2004, 'but I must confess I lack good arguments against its introduction, at least as an option for therapeutic reasons.'

We have been fooled by *Brave New World*, it seems: there is no coming dystopia wrought by anarchist scientists. Would anyone seriously regret leaving behind a world contaminated by birth defects, genetic disease, infertility, miscarriage and all the other hallmarks of natural reproduction? A full 75 per cent of natural conceptions ultimately end in failure – many before the woman even knows she has conceived. As Roger Gosden has recently said, in fifty years' time, where there is the will, the education and the resources, donor eggs and sperm will be regarded as an outdated, interim solution and virtually all babies will be born healthy. Who will thank those who stood against this progress? When the baby is born, and is in its mother's or father's arms, will there be any less wonder or excitement?

If our experience of IVF is anything to go by, there will not. We are already moving away from the naive sense that the wonder we feel at a new life comes from ignorance of the biology behind it: we are no longer ignorant, but we still have that sense of awe. As we take control of these processes, improving the lives of a vast number of humans, we will continue to enjoy the 'miracle' without experiencing the fear, the heartache and the disappointment. The only thing to be diminished by the approaching tide of reproductive technology is the assumed role of the divine in the creation of life. The secret anarchists are taking over from God.

And it is happening everywhere. On the frontiers of zoology, for instance, humans are being stripped of their status as 'special' animals. It is almost a given of most world religions that humans stand above the rest of nature; in the Bible's Book of Genesis, God gives Man 'dominion over the fish of the sea, and over the fowl of the air, and over the cattle, and over all the earth, and over every creeping thing that creepeth upon the earth'. However, geneticists have shown that the differences between us and other animals are only slight. So far we have identified only three genes that are unique to humans: the rest we share with our compatriots in the animal kingdom. When the catalogue of human genes is complete, the likelihood is that fewer than twenty of our 20,000 or so genes are unique to humans. We have also discovered that other primates have brain cells exactly like those inside our own oversized skulls. It's no surprise, then, that our seemingly unique mental capacities are nothing more than sophisticated versions of tricks that other animals can pull off. In a world where killer whales and dolphins show distinct cultural groups, crows use tools, chimpanzees display morality, elephants show empathy, and even salamanders and spiders show a range of personalities, it is hard to argue that there is anything biologically special about humans. It is true that nothing in the animal kingdom is using what we call language, but gestures used by bonobos and orang-utans come close.

Science is closing the gap between humans and other animals so fast, in fact, that some scientists are starting to think that human rights should also apply to other primates. In 1993, a team of eminent scientists published a collection of essays aimed at persuading the United Nations to grant chimps and other higher primates the privileges we know as 'human rights'. *The Great Ape Project* demanded that 'non-human hominids' – chimpanzees, gorillas, orang-utans and bonobos – should enjoy the right to life, freedom and protection from torture. The book included contributions from primatologist Peter Singer and evolutionary biologist Richard Dawkins, who writes in typically witty and provocative style:

> Remember the song, 'I've danced with a man who's danced with a girl, who danced with the Prince of Wales?' We can't quite interbreed with modern chimpanzees, but we'd need only a handful of intermediate types to be able to sing: 'I've bred with a man, who's bred with a girl, who's bred with a chimpanzee.'

Humans, Dawkins says, are not far enough removed from chimpanzees for there to be distinctions between their rights and ours. Stumble across one of those missing intermediates in the wild, he says,

> and our precious systems of norms and ethics would come crashing about our ears. The boundaries with which we segregate our world would be all shot to pieces. Racism would blur with speciesism in obdurate and vicious confusion. Apartheid, for those that believe in it, would assume a new and perhaps a more urgent import.

Predictably, reactions have varied from disbelief to scathing contempt for the idea. The Catholic Church condemned the Project for eroding the Biblical hierarchy that gives humans dominion over the Earth. Fernando Sebastián, the archbishop of

Pamplona and Tudela, called the idea ridiculous. 'We don't give rights to some people – unborn children, human embryos, and we are going to give them to apes,' he complained to BBC News when Spain said it would consider giving primates equal status to humans.

If the issues seem radical, then it is because biology has only quite recently started to challenge our taboos and move us out of our comfort zones. Physics, though, has been a thorn in humanity's side for hundreds of years.

At the beginning of the sixteenth century, the Earth sat at the centre of the universe, and according to received wisdom everything in the cosmos revolved around our planet. More than a thousand years earlier, the Egyptian astronomer Ptolemy had set out an intricate and beautiful (and mathematically complex) system describing the planetary orbits. By the end of the sixteenth century, however, the whole edifice was crumbling in the face of Nicolaus Copernicus's heliocentric cosmology.

In the light of what we have learned so far in our exploration of scientific anarchy, it is interesting to find that Copernicus's source was as strange and irrational as Einstein's: Copernicus was inspired by the bizarre ideas of a little-known Greek mystic.

Philolaus of Croton was a contemporary of Socrates. The cosmos came into being, he said, as a result of a complex arrangement of the primary elements, known as 'limiters' and 'unlimiteds'. The unlimiteds included earth, air, fire and water; limiters were shapes, such as tetrahedra, that fitted mathematically with the unlimiteds. From all this, Philolaus concluded that the Earth orbits a 'central fire'. And he wasn't talking about the Sun.

Philolaus' central fire is a mythic, religious flame that gave birth to the universe. To the Greeks it was also known as the Hearth of the Universe and the Watchtower of Zeus. In their view, the heavenly bodies orbited this central fire in ten concentric circles. Farthest out were the 'fixed' stars, then the five (known) planets, then the Sun, the Moon and the Earth. In the innermost orbit was the 'counter-Earth'. This body remains invisible to us, because the Earth rotates on its axis as it revolves around the central fire; we always have our back to the counter-Earth, Philolaus solemnly pronounced.

In his defining work, *De Revolutionibus*, Copernicus cites this system as a 'precursor' to his astronomical system. Copernicus did not build on the accumulating scientific knowledge of the previous decades and centuries, but jumped back to the time of the mystical Greeks. This is rarely, if ever, mentioned by modern astronomers, for whom Copernicus is the patron saint of reason. This is simply not how things are supposed to be. Years after *De Revolutionibus* was published, Galileo expressed his shock. Contemplating how Copernicus took Philolaus' strange and mystical ideas and turned them into something so self-evidently true, Galileo said in his *Dialogue Concerning the Two Chief Systems of the World*, 'there is no limit to my astonishment'. Ptolemy would have been turning in his grave: before his death in AD 168, Ptolemy referred to Philolaus' ideas as 'entirely ridiculous'.

Perhaps it was this source, acknowledged by Copernicus, that explains why the scientific establishment was not particularly convinced by Copernicus's argument that the Earth must go round the Sun. For instance, the astronomer Tycho Brahe, born shortly after Copernicus died, simply refused to believe it. More problematic, though, was the fact that it flatly contradicted the sacred view of how the universe held together.

That is certainly why Galileo's attempts to prove Copernicus right landed him in trouble; he ended his days under house arrest. Nevertheless, the anarchist Isaac Newton took up where Galileo had left off, and set the heliocentric universe on a firm mathematical footing. Newton left a little wiggle-room for God, though – he declared that there was still room for the divine hand to steer the fine details of planetary motion. Soon afterwards, though, Pierre-Simon Laplace made God redundant. His mathematics dotted the i's and crossed the t's; Laplace famously (and brazenly) declared that his calculations of the motion of the heavenly bodies were so accurate that there was now 'no need' for God's involvement.

Not that this was Laplace's explicit purpose. It is simply what science does: in pursuit of understanding it assaults the status quo and takes pity on neither God nor man. And as it began, so it carries on. Physicists have since taken their audacious theories all the way back to the beginning of everything. The Big Bang theory's description of how the universe came into being is breaking taboos even now. In September 2010 the front page of the London *Times* screamed out that Stephen Hawking had declared, Laplace-like, that God is defunct as Creator: the laws of physics could bring the universe into being without divine assistance. It was a strange pronouncement, and caused widespread outrage among religious commentators. But there was really nothing surprising about Hawking's statement – he already had form in this area.

In *A Brief History of Time*, Hawking tells the story of a 1981 conference on cosmology, held at the Vatican, which he attended. At one point during the proceedings he met Pope John Paul II, and the two men discussed the merits of cosmology. According to the Pope, it was all right to study the evolution of the universe after the Big Bang, but not the Big Bang itself: 'that was the moment of Creation and therefore the work of God,' said the Pope. With typical dry wit, Hawking gleefully tells his readers that he had just given a presentation to the conference on that very topic. He didn't tell the Pope, though. 'I had no desire to share the fate of Galileo,' he says.

Hawking is joking, obviously: he wasn't worried at all, because the secret anarchists have rendered the Papacy toothless. In fact, science has religion running scared. Hawking is nothing but amused, it seems. Read between the lines and he seems to be saying, 'Let the Pope have his fear of God: we scientists are afraid of nothing.'

Except, he could have added, one another. Scientists are happy to take occasional potshots at the Almighty, but it is much more common for them to assault their peers. As we will explore in the next chapter, science is a brutal, gladiatorial arena in which its anarchy finds extraordinary expression.

6

FIGHT CLUB

There's no prize for the runner-up

'I played over the music of that scoundrel Brahms. What a gift-less bastard. It annoys me that this jumping, inflated mediocrity is hailed as a genius.' That was Tchaikovsky's assessment of one of his more celebrated contemporaries. Such attitudes run through the history of the arts. Louis Spohr, a German violinist and composer, called Beethoven's Fifth Symphony an 'orgy of vulgar noises'. Édouard Manet wrote to his colleague Claude Monet about Renoir in less than glowing terms: 'He has no talent at all, that boy. Tell him to give up painting.'

The arts, though, are a sycophant's paradise compared with the sciences. At least the artists snipe about their colleagues behind their backs; scientists do it face to face. 'Our speaker today is a man about whom we have heard so much, and from whom we have seen so little.' That was the chemist Gilbert Lewis's introduction when Irving Langmuir visited his department to give a seminar. Four years later, on 23 March 1946, an hour or so after having lunch with Langmuir, Lewis was found dead in his laboratory. The air was filled with the scent of almonds – and a flask of hydrogen cyanide was open on the laboratory bench.

To this day, no one knows whether Lewis killed himself, whether it was an accident, or whether something more disturbing took place. No autopsy was ever carried out on Lewis's body. In *Cathedrals of Science*, a fascinating dissection of the history of chemistry, Patrick Coffey uncovers a trail of lies and half-truths linked to the incident. Langmuir's visit to Berkeley that day was conveniently airbrushed out of the story for nearly sixty years. In Langmuir's writings, dates were fudged: he visited the University of California in '1945 or 1946' according to his celebrated 'Pathological Science' essay in which he takes Lewis's science to pieces. Joel Hildebrand, who organised the lunch, later wrote – erroneously – that it was in 1945. Whether that was a subconscious slip or a deliberate attempt to remove Langmuir from the vicinity of Berkeley at the time of Lewis's death, we will never know. Everyone involved is now dead.

Coffey concludes that Lewis most probably had a heart attack in his laboratory while he was working with hydrogen cyanide. He had led an unhealthy, tobacco-fuelled, hate-filled life and was a prime candidate for heart failure. But he was also depressed, and others felt that suicide was the most likely scenario. Murder is inconceivable. Or at least no one seems to mention it as a possibility.

Anyway, it was not necessary: in scientific terms, Langmuir had already killed Lewis. Langmuir had won a Nobel Prize for work that Lewis considered his own. Lewis had sown the seeds of his own destruction, though, by making bitter enemies of two chemists who were highly influential in Stockholm. Walther

Nernst and Svante Arrhenius hated each other; Arrhenius had managed to block Nernst's much-deserved Nobel Prize for fifteen years. But Lewis gave them a common enemy. In a 1907 paper he had called some of their work 'unsystematic' and 'inexact'. Their 'old approximate equations' would 'no longer suffice'. The paper rendered Lewis's hopes of a Nobel Prize as good as dead.

In October 2010, a study by Dutch researchers found that school-children who are timid and introverted are more likely to go into science – like lambs to the slaughter, one might say. Though few would openly admit the fact until recently, it is now clear that if you want to achieve greatness in science, you need to be ready to kill or be killed. In the race to discovery, there are no prizes for second place. As Peter Medawar wrote:

> Much of a scientist's pride and sense of accomplishment turns ... upon being the *first* to do something – upon being the man who did actually speed up or redirect the flow of thought and the growth of understanding ... Artists are not troubled by matters of priority, but Wagner would certainly not have spent twenty years on *The Ring* if he had thought it at all possible for someone else to nip in ahead of him with *Götterdämmerung*.

This is a theme that is repeated throughout science – especially where Nobel Prizes are involved. Take the story of the transistor, for example. The transistor is the defining technology of the modern world. Its function sounds prosaic: it is, essentially, a switch and amplifier for electrical signals. But your life would not function without transistors. Besides their obvious uses in computers, mobile phones and internet servers, they sit inside toasters and washing machines, cars and microwave ovens. Take away transistors, and today's world would be unrecognisable.

In the twenty-first century we have learned to make transistors almost unimaginably small: in 2008 Bell Labs unveiled one made from a single molecule. The world's first transistor, by comparison, was around a centimetre tall. It, too, was born at Bell Labs. And it was born out of anarchy: Walter Brattain, John Bardeen and their boss William Shockley won the 1956 Nobel Prize in Physics for its invention, but by the time they arrived in Stockholm they were bitter enemies.

The conflict began in 1947. Brattain remembers Shockley bursting into the laboratory where he and Bardeen were working on a design for an amplifier that would be the forerunner of the first transistor. Shockley had learned that they had been talking with Bell Labs' patent lawyers, and he reminded them that it was he who had shown how to control the electric current flowing through a piece of silicon: any patent should be his, he warned them. 'Oh hell, Shockley,' was Brattain's reply. 'There's enough glory in this for everybody!'

Shockley was unappeased. He decided to go it alone, and took his own transistor design to the patent lawyers. It was a disaster. Their search for precedents uncovered the fact that, in 1930, the physicist Julius Lilienfeld had filed a US patent for a transistor identical to Shockley's. The only way Bell Labs would be able to patent a transistor, the lawyers told him, was if they used Bardeen and Brattain's design.

For a month, Shockley agitated, sleeping badly at night and scheming by day. Then he had what Michael Riordan and Lillian Hoddeson, the authors of *Crystal Fire*, call 'the most important idea of his life'. It was a sandwich of semiconducting materials that would amplify electrical signals in a whole new way. He told Bardeen and Brattain nothing about it, and worked on the design in secret.

Three weeks later, one of their Bell Labs colleagues unwittingly forced Shockley's hand. John Shive had had a similar idea, and mentioned it in a seminar talk. Shockley, worried that Bardeen and Brattain would quickly fill in the gaps and reinvent his amplifier, got to his feet and upstaged Shive's talk with a complete description of his new design. Bardeen and Brattain were dumbfounded to discover that their boss had been keeping such a well-developed idea from them. With nothing more than an exchange of glances, war was declared.

Bardeen and Brattain had a working device, and could accelerate their patent application. Shockley had nothing to experiment with, but he did have one thing his enemies didn't: authority. Within a few days, Bardeen and Brattain found that all the labs' resources had been redirected to focus on developing Shockley's device.

It was too late, however. Bardeen and Brattain's invention worked. They refined and renamed it, ready for launch, as the 'transistor'. All Shockley could do was tack a supplement onto their work. The transistor was launched in three papers in the 15 July 1948 issue of *Physical Review*. Two were, as Riordan and Hoddeson put it, 'ageless classics' written by Bardeen and Brattain. The third was a 'largely forgettable' paper by Shockley and an experimental collaborator, Gerald Pearson.

Shockley was down, but he was not beaten. Having failed to halt the birth of the transistor, he resolved to find another way to write himself into the history books. He called in favours with his managers, and mounted a PR offensive that would make Stanley Prusiner blush. Astonishing as it now seems, Shockley managed to get his friends in the upper echelons of Bell Labs to insist that no photographs of Bardeen and Brattain were to be taken without Shockley in the frame. At the press conference to launch the transistor, Bardeen and Brattain were completely sidelined. Shockley's boss, Ralph Bown, made the announcement. When his demonstration of the transistor's power was over, Bown handed the floor to Shockley, who fielded the press's questions.

The evidence of Shockley's anarchy is still there for anyone to see. The classic Bell Labs photograph of the trio at work shows Shockley sitting at the laboratory bench making adjustments to the transistor while looking through a microscope. Bardeen and Brattain are standing behind him, watching. The pained expressions on their faces are of men worried that their most precious possession is about to be broken by an inexpert hand. Bardeen, a man of very few words (but the holder of two Nobel Prizes), once let slip that Walter Brattain 'sure hates this picture'.

And it wasn't over even then. Shockley had dealt with the transistor affair, but he wasn't going to risk letting something similar happen again. He began a campaign of systematic exclusion, creating what Bardeen later called 'an intolerable situation'. When Shockley's research team moved to a new building, Bardeen and Brattain were allocated space on the floor beneath Shockley and his close collaborators. Complaints to Bown got them nowhere. By 1950, six years before the trio were awarded their Nobel Prize, Bardeen had given up. He moved into a different field entirely – research into the electrical properties of materials known as superconductors – and left Bell Labs a year later. 'My difficulties stem from the invention of the transistor,' he wrote in his resignation letter to the research director. Walter Brattain stayed put but, sore at the loss of his friend and collaborator, he refused to work with Shockley ever again.

William Shockley is not an unusual type of character to find in science. He was a great scientist, but an even greater combatant. When he died in 1989, it was recorded in Stanford University's memorial notice that 'He set extraordinarily high standards for everyone, including himself, and virtually every activity was an all-out race in which he was an intense competitor.' It seems that when he was losing that race, he wasn't above bending some of the rules so that he could muscle in and share the glory. But Shockley's behaviour is only the thin end of the wedge when it comes to scientific infighting. Stealing the limelight is nothing: real anarchists stubbornly refuse to acknowledge that there is even anything worth stealing. And then they convince their colleagues of it too.

Though scientists like to hold up Copernicus as a researcher who was obviously right, his golden idea – that the Earth goes round the Sun – was widely rejected by his scientific peers. Tycho Brahe, a giant of astronomy in the years after Copernicus's death, was among those who chose to ignore the heliocentric model. Then there are Isaac Newton and Friedrich Gauss, who both waited twenty years for recognition and acceptance of their radical ideas. A full thirty-five years passed before Newton's own university was willing to teach his work.

Jonathan Swift once said that 'When a true genius appears in this world, you may know him by this sign, that the dunces are all in confederacy against him.' That was certainly the case for Alfred Wegener, the man who proposed a theory of continental drift decades before we had any understanding of plate tectonics. Wegener had noted that various coastlines – notably the Atlantic coasts of South America and Africa – would fit together well if not separated by the ocean. The geology also seemed to span the water: mountain ranges and other features would continue across the divide. While reading in a library in Marburg, Germany, Wegener had come across the curious fact that the fossil records of South America and Africa shared common specimens. The evidence was not conclusive, but, taken all together, it looked persuasive. 'A conviction of the fundamental soundness of the idea took root in my mind,' he later wrote.

In 1912, Wegener presented his idea to a meeting of the Geological Association in Frankfurt. He proposed that in the distant past the Earth's continents had been joined together in a supercontinent, later named Pangaea. The continents had subsequently 'drifted' apart. The idea was summarily dismissed by geologists. In 1926, the American Association of Petroleum Geologists even organised a special symposium to denounce Wegener's hypothesis. It had a lasting effect: until the early 1960s, continental drift was regarded as a crazy idea. Then the right technology arrived in scientists' hands, and Wegener was exonerated.

During the Second World War, the Allies had constructed magnetic maps of the sea floor in an effort to track German submarines. By the 1960s, scientists were able to use those maps – and the mapping equipment – to examine the nature of the Earth's crust. They discovered that the Earth's crust consists of a jigsaw of huge interlocking plates, all riding on the semi-molten layer below, and slowly moving in relation to one another. At a meeting of the Royal Society in 1964, scientists pronounced continental drift to be the new orthodoxy. By the mid-1960s, you could not get a paper published that did not embrace Wegener's hypothesis. Sadly, by then Wegener had been dead for more than thirty years.

John James Waterston's recognition was similarly slow and posthumous. In 1843 he came up with a description of the behaviour of gases that was a forerunner of the now standard kinetic theory of gases. When he submitted it to the Royal Society for peer review, Sir John Lubock described it as 'nothing but nonsense'. It was forty-five years before Waterston's contribution was recognised.

Sometimes, though, decades spent in the wilderness being scorned and insulted by colleagues are eventually rewarded.

Barbara McClintock, for example, had been dismissed by one prominent geneticist as 'just an old bag who'd been hanging around Cold Spring Harbor for years'. How sweet it must have been, then, when she won a Nobel Prize in 1983 in recognition of work she had carried out – and been utterly derided for – nearly forty years earlier.

McClintock's discovery is reflected in newspaper headlines across the world. If we had paid attention to it earlier, our hospitals may not have had to deal with deadly 'superbugs', the rapidly evolving bacteria that outcompete our best antibiotics. The antibiotic-resistant Staphylococcus bacterium we know as MRSA, for instance, contributed to the death of around 13,000 Britons between 1993 and 2009 – and we are only just getting its impact under control. These killers are so effective because of an evolutionary mechanism that, according to McClintock's peers, simply couldn't exist.

You get a hint of McClintock's steely resolve in the face of such criticism when you hear the story behind her name. Her parents initially called her Eleanor, but decided that was too delicate and feminine. 'Barbara' McClintock encapsulated her temperament much better, apparently. Perhaps it was because St Barbara was the protector against lightning strikes and firestorms, or simply because the name means 'outsider' or 'foreigner'. How her parents had such foresight, it is hard to know.

In December 1941, McClintock began working at the Carnegie Institution of Washington's research facility on Long Island, New York. The laboratory, a genetics research station where plant breeders tried to find the roots of heredity, was called Cold Spring Harbor. McClintock, now thirty-nine years old, had been offered a one-year research post, but she was destined to enter her fifth and sixth decades still working for the Institution. Until this point she had flitted around, working in Germany, in California, in Missouri and at Cornell. On Long Island, though, she found what she had been searching for: the chance to immerse herself fully in discovery.

McClintock's speciality was the genetics – the hereditary characteristics – of maize. Just as you and I have a certain hair and eye colour, or fingernail shape, that we get from our parents, maize plants have characteristics, notably the colour of leaves and kernels, that are determined by their parent plants. McClintock grew hundreds of maize plants at a time and kept track of their parentage. In fact, she did more than that: she sidelined the role of wind and insects, and pollinated them by hand.

Each kernel of corn comes from a single egg, one of thousands on the plant. The fertilising pollen comes either from the same plant, or from a neighbouring plant and carried across by the wind or by insects. Each egg can thus be fertilised from an entirely separate plant. This is a recipe for a huge amount of variation within a single organism. For a geneticist, that is both a blessing and a curse: the variation gives the potential for interesting traits to be selected for. But with so many variables, it is easy to lose track of the data so vital to a scientific understanding.

McClintock, though, was thorough, passionate and single-minded. Though she regarded many colleagues as friends, she was uninterested in personal attachments beyond friendship. She once told her biographer, Evelyn Fox Keller, that 'I just didn't feel it. And I could never understand marriage.' She was, in many ways, entirely suited to this lonely task, heading out into her Long Island maize fields in the early morning to focus all her attention on kernels of corn, the waxy striped leaves, and the molecular structures within the plant that gave rise to all its characteristics.

In 1944, McClintock noticed something odd in a plant she had labelled B-87. Its yellow kernels were speckled with red and purple spots. There was nothing intrinsically peculiar in this; it was the way the spots had appeared that caught McClintock's attention. Each kernel starts out as a single cell that repeatedly divides. With each division, a copy of the genetic recipe for the cell is passed on. Because every cell has the same recipe, its colour – the default is yellow – should be consistent. If it isn't, a gene for colouring the cell with a pigment – turning it purple or red, for instance – must be turning on sporadically. What McClintock's keen eye noticed in B-87 was that, though the pigmentation was sporadic, it was far from random.

The standard theory of inheritance had two problems with this. First, a gene might spontaneously turn on, causing a mutation, but it was then meant to stay turned on. A mutation was thought to be a permanent feature. Second, the mutations were meant to be random. With B-87, this was definitely not the case. When McClintock examined how many coloured spots there were of each particular size, there was a definite pattern. When she looked at the plant as a whole, watching it grow new kernels, that pattern was repeated.

Large spots were produced when a pigmentation gene turned on early in the cell division process, giving time for the colourful mutation to be repeated. Small spots came from a mutation appearing late in the growth cycle of the kernel. And the relative prevalence of large, small and medium-sized spots made it very clear that this was not a random process. Something was controlling the mutations – but McClintock knew that this was biological heresy. It took three years of further breeding, observation and examination of the cells under the microscope before she eventually found a reason to believe that she wasn't fooling herself.

Years of studying the speckling of kernels had led McClintock to study the variegation, or stripiness, of the plant's leaves. In 1946 she had noticed that there was a pattern to variations in the variegation: if there were ten stripes per centimetre on average, she would often find that a region of the leaf that was two stripes above the average sat next to a region that was two stripes below average. To McClintock, it looked as if one cell had gained what an adjacent cell had lost. Somewhere along the line from gene to cell to leaf, something had been passed from one thing to another – and in a tightly controlled way. The agent, she eventually found, was in the chromosomes.

The science writer Matt Ridley has an interesting way to describe chromosomes. Chromosomes are made of DNA, and the fundamental units of DNA are four molecules known as bases – adenine, cytosine, guanine and thymine – which we represent by the letters A, T, C and G. These letters can be strung together to make words that form paragraphs that will eventually form a chapter. That chapter, Ridley suggests, is the chromosome. Put all the chromosome chapters together, and you have a book that perfectly describes the instructions for making an organism.

You and I are described by twenty-three pairs of chromosomes in our cells, one of each pair coming from each parent. Maize cells contain ten pairs of chromosomes. McClintock's heresy took place on chromosome nine. She found that entire paragraphs were moving around within this chapter. She called the process transposition, and suggested that this is what causes the variegation in the leaves and the speckle in the kernels. The analogy is not strictly correct, but it is as if the 'reader' of the genome instruction manual was occasionally coming across an instruction to turn on a purple pigmentation just before the paragraph that told it to stick with the default yellow, whatever it might read later.

Even today, biologists don't fully understand transposition, but by 1951 McClintock was able to publish a stab at what was going on inside the chromosome. Parts of the genome would move around in a co-ordinated fashion, she said, vastly increasing its repertoire of products. These 'controlling elements' were not mere genes; they were more like managers put in charge of the genes. The rigid recipe book of the genome had turned out to be more than a list of ingredients: it was a complete kitchen, equipped with creative chefs, ingredients and utensils. And the controlling elements would manage the kitchen, determining how all this potential was to be put to work.

To McClintock, the seemingly endless variation in the biological world suddenly made a lot more sense. Biologists had struggled to understand how, when cells in the organism had exactly the same set of genes, some would turn into muscle while others created arteries or produced a filament of hair. McClintock had an answer: through the action of the controlling elements. However, her colleagues, for the most part, simply didn't believe her.

There is still great controversy over whether McClintock was wronged by her colleagues. Some historians of science have suggested that she was a woman in a man's world, and that the rejection of her ideas was sexist in origin. Others say that the evidence simply wasn't compelling enough for such radical conclusions to be drawn. Judging by a letter she wrote in 1973, McClintock felt that she was simply ahead of her time: 'This became painfully evident to me in my attempts during the 1950s to convince geneticists that the action of genes had to be and was controlled ... One must await the right time for conceptual change.' Another letter to a colleague, written the same year, betrays more anger: 'I stopped publishing detailed reports long ago when I realized, and acutely, the extent of disinterest and lack of confidence in the conclusions I was drawing.'

It is interesting that decades passed before McClintock began to comment on the reception her work had received. At the time, it was simply a delicious reprieve – it meant she was left alone.

The Nobel Prize-winning biologist Albert Szent-Györgyi, the discoverer of vitamin C, seems to have understood the independent streak that runs through science. 'The real scientist ... is ready to bear privation and, if need be, starvation rather than let anyone dictate to him which direction his work must take,' he said. But make no mistake, he pointed out, this is not a selfless pursuit of truth, but a selfish one: 'Research uses real egotists who seek their own pleasure and satisfaction, but find it in solving the puzzles of nature.' Szent-Györgyi could have been writing about Barbara McClintock.

McClintock joined the Cold Spring Harbor Laboratory because it offered her a 'policy of no interference and complete freedom'. 'I just go my own pace here,' she said, 'with no obligations other than that which my conscience dictates.' We can see her egotism in her disregard for informing anyone else about her work. 'I decided it was useless to add weight to the biologist's wastebasket,' she wrote in a letter to a fellow geneticist in 1973. 'Instead, I decided to use the added time to enlarge experiments and thus increase my comprehensions of the basic phenomena.'

This approach freed her from the criticisms of colleagues, and let her set her own research agenda and make her own corrections. She said in 1983:

> Over the many years, I truly enjoyed not being required
> to defend my interpretations. I could just work with the
> greatest of pleasure. I never felt the need nor the desire to
> defend my views. If I turned out to be wrong, I just forgot
> that I ever held such a view. It didn't matter.

McClintock, it seems, was not in science for recognition: her reward was the private satisfaction of the puzzle-solver. She was in science simply because she wanted to know – for herself. In fact, we get no sense from her writings that she felt particularly slighted when, in 1965, the French geneticists François Jacob, André Lwoff and Jacques Monod won their Nobel Prize; in 1960, the trio had shown that genes for building body parts were flanked by genes for regulating them.

Others, though, were offended on McClintock's behalf. In 1967, when she was awarded a National Academy of Sciences Kimber Medal, the citation pointed out that McClintock's work was the precursor to the Nobel laureates' work, and that 'their thinking was probably much influenced by Barbara's notion'.

Suddenly – and probably, to McClintock, annoyingly – she was back in vogue. Transposition was fingered in antibiotic resistance, cancer and immunology. It was universal in nature, not just a quirk of maize. In 1981 McClintock was, in the words of one of her biographers, 'besieged with awards': five prestigious prizes came her way. Two years later, in 1983, she was in Stockholm wondering what to say in her Nobel acceptance speech.

Before the award ceremony, Nobel laureates stay in Stockholm's Grand Hotel. Its views across the waterfront to the magnificent Royal Palace are postcard-perfect, with boats moving through the waterways and tourists strolling along the embankments, all set against a backdrop of the finest Swedish architecture. It was perhaps while she was sitting at a desk by a window overlooking such a view that Barbara McClintock contemplated revenge on her colleagues.

The first drafts of her acceptance speech, jotted down on the hotel's headed notepaper, certainly give that impression. The note-paper is a simple design, with just a crown – the hotel's logo – and a few addresses and telephone numbers. The twenty pages of notes that McClintock made while preparing her speech are more complicated, however. There are crossings-out, and meandering amendments that curve and swoop around the pages. Her handwriting, like her science, is not easy to decipher. Nevertheless, it is worth decoding because it provides a rare insight into the mind of a scientist who has climbed the biggest peak, yet become an object of ridicule, scorn and callous snubs.

'For many years,' she began, 'I worked on a genetic phenomenon that was most unacceptable to all but a few persons.' Then she crossed this out as a false start. Below it, she wrote that she was 'truly amused' by some of her critics. A maize geneticist came to see her once, and said he had heard that she held some strange views, and that he did not want to hear a word about them. 'I could not refrain from laughing' she said.

That story is outlined in almost every draft of the speech, but it didn't make the final cut. At the ceremony, McClintock contented herself with expressing pleasure at her 'radical' status. Because her peers and colleagues ignored and even rejected her, she had been left alone to get on with the work she loved:

> I was not invited to give lectures or seminars, except on rare occasions, or to serve on committees or panels, or to perform other scientists' duties. Instead of causing personal difficulties, this long interval proved to be a delight. It allowed complete freedom to continue investigations without interruption.

The period in which she was isolated 'surprised and then puzzled' her, she said, but she didn't mention how her colleagues' blindness amused her, and how she laughed out loud at their more ridiculous reactions to her work. Instead, she reiterated her anarchic selfishness. For her, the work was always about the puzzles of genetics and the 'pure joy' they provided:

When you suddenly see the problem, something happens
– you have the answer before you are able to put it into
words. It is all done subconsciously. This has happened too
many times to me, and I know when to take it seriously.
I'm so absolutely sure. I don't talk about it, I don't have to
tell anybody about it, I'm just sure this is it.

It seems that McClintock shared the sentiment expressed by the French physiologist Claude Bernard: 'The joy of discovery is certainly the liveliest that the mind of man can ever feel.'

Not all scientists can rise above the pain of rejection, though. Svante Arrhenius, the first Swedish recipient of the Nobel Prize, moved heaven and earth to make sure that his genius was recognised. As a graduate of the Cathedral School in Uppsala, he was no doubt aware of the biblical principle that 'a prophet hath no honour in his own country'. He had carried out a groundbreaking analysis of chemical reactions while still a postgraduate student, but his superiors considered the work unremarkable, and he was given the lowest possible pass. That would have scuppered his chances of a career, but Arrhenius won himself scientific salvation by sending his thesis to prominent scientists outside Sweden. The foreign physicists and chemists recognised its quality – it was, after all, the same work that would later bring him his Nobel Prize – and showered him with offers of research positions. Keen to stay in Sweden to care for his dying father, Arrhenius used the offers as leverage to broker himself a job in Uppsala.

It was an early indication of Arrhenius's taste for the Machiavellian. In a delicious piece of anarchic manoeuvring, for instance, he sowed the seeds of his own Nobel Prize. He was asked to help set up the Nobel Foundation; knowing that he stood little chance of benefiting from Alfred Nobel's legacy if his own countrymen had sole charge, Arrhenius used his position to ensure that the nomination committee would include scientists from outside Sweden.

It is easy to romanticise the early years of the Nobel Prizes as a time when science was still finding its way in the world. Perhaps the intrigue, insults, infighting and machinations at the Karolinska Institute in the first half of the twentieth century were just outward signs of an adolescent phase that science was going through. It's a nice thought, but it doesn't stand much scrutiny. As the case of Barbara McClintock showed, scientists' predilection for conflict, insult and disparagement of peers remained strong through the post-war period. And things aren't much different now.

In July 2010, Nobel manoeuvrings came to the fore over a prize that may never be awarded. The Higgs boson, a particle that our best theories of physics suggest is what endows objects with mass, has yet to be discovered. But five people are in line to receive the Nobel Prize in Physics should the discovery be made. And, since a Nobel can be shared between three scientists at most, blood will be spilt.

The problem is one of priority. The British researcher Peter Higgs's name is associated with the particle only by accident. In 1967, Higgs shared a bottle of wine with a US researcher called Benjamin Lee. In 1972, Lee was the rapporteur for a conference at the US National Accelerator Laboratory (now Fermilab) in Batavia, Illinois, and used Higgs's name as a shorthand for the idea. The name stuck.

The existence of the Higgs boson was actually suggested by three different research groups within a few weeks of each other. Two Belgians, Robert Brout (who died in May 2011) and François Englert, were first to the idea, in August 1964. Then, a fortnight later, came Higgs's brief paper about it in *Physics Letters*. By the time a three-way UK–US collaboration published their paper four weeks after that, all the seats in Stockholm – should they ever be made available – were taken. One of the third-placed contributors expressed his disappointment in a paper published in 2009: 'We were naive enough to feel that these other articles offered no threat to our insights or to the crediting of our contribution. Nearly 45 years later, it is clear that we were very wrong.'

The disappointment simmered, then boiled over when the announcement of a meeting in Paris to discuss the latest results in the hunt for the Higgs boson mentioned only the first three scientists. In a reaction that belied scientists' reputation for a calm, measured, patient outlook on life, a number of particle physicists, mostly US-based, threatened to boycott the meeting or stage a protest.

'Anyone who witnesses the advance of science first-hand sees an intensely personal undertaking,' Carl Sagan once wrote. 'A few saintly personalities stand out amidst a roiling sea of jealousies, ambition, backbiting, suppression of dissent, and absurd conceits. In some fields, highly productive fields, such behavior is almost the norm.'

Sagan might have been writing about the travails of his first wife, Lynn Margulis. Those who can see the impact of Margulis's work find it astonishing that she hasn't yet been awarded a Nobel Prize. Put simply, she suggested that much of our biology – the complexity of our cells, for instance – arose through two or more organisms co-operating together for mutual advantage. The idea is now known as endosymbiosis, and it is taught in every university biology department in the world.

Perhaps Margulis has her own personal Arrhenius blocking the nominations. It's certainly not because her idea isn't big enough – that much is clear from reading what other scientists say about it. The American palaeontologist Niles Eldredge calls it 'probably the grandest idea in modern biology', while philosopher of science Daniel Dennett says that it is 'one of the most beautiful ideas I've ever encountered'. For Richard Dawkins, it is 'one of the great achievements of twentieth-century evolutionary biology'. No wonder, then, that the Chilean biologist Francisco Varela referred to Margulis as 'one of the brightest and most important biologists since the geneticists of the 1920s'.

These accolades didn't come without a fight – a fight, in fact, that Margulis waged against many of the people who now acknowledge her brilliance. She is a textbook anarchist. She is, Dawkins says, 'extremely obstinate ... the kind of person who just knows she's right and doesn't listen to argument'. Engineer and inventor Daniel Hillis is more direct about the anarchy. To most biologists' minds, he says, Margulis 'didn't follow the rules and pissed a lot of people off'.

But she had to. 'Most of the science that gets done gets done within a rigid set of rules, where you know exactly who your peers are, and things get evaluated according to a very strict set of standards,' says Hillis. 'That works, when you're not trying to change the structure ... when you try to change the structure, that system doesn't work very well.' He is right. Everyone else who had tried to get this idea across played by the rules – and without success.

In 1883, the French botanist Andreas Schimper made an interesting observation. He saw that chloroplasts, the parts of plant cells that make energy from sunlight, divide in exactly the same way as cyanobacteria, which also make energy from sunlight. Perhaps, Schimper tentatively suggested, this was because green plants were the result of some kind of merger between other biological organisms.

Konstantin Mereschkowski, who lived and worked in Kazan, Russia, liked this idea. He explored it in his herbarium, where he kept more than 2,000 specimens of lichen. Many of Mereschkowski's lichen were 'symbiotic' combinations of fungi and cyanobacteria. In symbiosis, two organisms benefit from the other's presence. In some lichens, for example, the bacteria gain a water and mineral trap, and the fungus gains access to nitrogen that the bacteria can pull from the air. Around one-fifth of fungi are now known to live in such a symbiotic relationship.

In 1905, Mereschkowski proposed that biological complexity might have arisen from such arrangements being made permanent. If simpler organisms start to live symbiotically, would it be such a stretch to imagine them becoming biologically unified? Yes, was the reply from biologists – and yes it remained for more than sixty years. As the ideas of Darwinian evolution took a firm hold, it became established dogma that biological change happened through the slow evolution of one species into another. Anyone who suggested that it could happen through a union of two organisms had to be deluded.

No wonder, then, that Ivan Wallin gave up on his research. In the 1920s, Wallin, who was working at the University of Colorado medical school, suggested that the mitochondria that generate energy in every cell in your body are actually enslaved bacteria. He made the suggestion because, looking down his microscope, he couldn't tell mitochondria and bacteria apart. After none of his nine papers on the subject were taken seriously, he gave up research and became a teacher and administrator. It was only when a true anarchist arrived on the scene that Mereschkowski and Wallin's ideas found the champion they needed.

As with Stanley Prusiner's prions, Lynn Margulis's big idea was essentially one that other people had had before. In her case it was Mereschkowski and Wallin's suggestion that the mechanics of life going on inside our bodies is the result of symbiotic relationships between bacteria and other organisms.

The idea is so hard for Darwin's successors to accept because it denies that all the variation in nature results from random genetic mutations. According to the classic genetics-plus-natural selection camp of biologists – known as neo-Darwinists – environmental factors produce mutations in the genome that are passed on to the next generation. Some mutations will manifest themselves as useful new characteristics which enable that generation to survive better than its neighbours. Other mutations will cause problems, and will die out.

According to Margulis, however, the most important variation that is passed on through the generations is what happens when plant and animal cells play host to microbial genes. This, she says, is the origin of complex life:

> It may have started when one sort of squirming bacterium invaded another – seeking food, of course. But certain invasions evolved into truces; associations once ferocious became benign. When swimming bacterial would-be invaders took up residence inside their sluggish hosts, this joining of forces created a new whole that was, in effect, far greater than the sum of its parts: faster swimmers capable of moving large numbers of genes evolved. Some of these newcomers were uniquely competent in the evolutionary struggle. Further bacterial associations were added on, as the modern cell evolved.

We now know, from the fossil record, that Margulis is right. There is plenty of evidence that such things went on all those millions of years ago. That is why, to use Dawkins' words, endosymbiosis has gone 'from being an unorthodoxy to an orthodoxy'.

It was the most difficult of journeys, however. The original paper containing this revolutionary idea was rejected fifteen times before it was finally published. Even on publication, those who didn't ridicule the idea ignored it or dismissed it as unimportant.

Faced with such scorn, Margulis decided to write a book to explain her ideas in full. This, Hillis says, was her most serious scientific crime: she sidestepped the journals' peer review system. 'In the minds of many people, she went around the powers that be and took her theories directly to the public,' Hillis says. This annoyed the biologists greatly – especially when she turned out to be on to something. 'If it's a sin to take your theories to the public, then it is a double sin to take your theories to the public and be right.'

Not that the book had an easy ride. Academic Press gave Margulis a contract, but when their peer reviewers poured scorn on the book's central ideas they refused to publish it. Eventually, Yale University Press took it on. When the book came out, in 1970, Margulis found that crime against her peers didn't pay. Suddenly she was *persona non grata*.

Here is how Margulis describes, in a letter to *The Sciences*, the reaction to a request she made for funds from the National Science Foundation grants for further work on endosymbiosis theory:

> I was told by an NSF grants officer (after having been supported nicely for several years) that 'important' scientists did not like the theory presented in a book I had written and that they would never fund my work. I was actually told that I should never apply again to the cell biology group at NSF.

Unlike Mereschkowski and Wallin, she didn't take rejection lying down. A flavour of Margulis's fighting spirit comes through in her contribution to John Brockman's book *The Third Culture*. She describes the great evolutionary biologist John Maynard Smith as an engineer who 'knows much of his biology second hand'. He and his fellow neo-Darwinists Dawkins, Eldredge, Richard Lewontin and Stephen Jay Gould 'codify an incredible ignorance'. Their work is 'reminiscent of phrenology' and 'will look ridiculous in retrospect, because it is ridiculous'.

Margulis does not pull her punches. But that is almost certainly why she is where she is today. She fought hard to get her idea accepted, and it wouldn't have happened if she had been a shrinking violet. She had to be an anarchist.

In the preface to her rule-breaking book, Margulis quotes the American geneticist Carl Clarence Lindegren. The scientific establishment, Lindegren declared, 'is permeated with opinions which pass for valid scientific inductions and with contradictions which are disregarded because it is too painful to face the prospect of the revisions of the theory which would be required to reconcile the contradictory observations with the dominant theory'. Margulis agrees: 'The problem is basically one of attitude and training,' she says. 'In trying to explain the central concept of the theory to various interested and helpful colleagues over the years, I found it necessary first to overcome this often inexplicit but powerful resistance to a new point of view.'

With typical brazen spirit, Margulis later accused her detractors of laziness in their search for evidence. Why study only what is currently alive, she asks, when the fossil record gives us millions of years of data? 'Richard Dawkins, John Maynard Smith, George Williams, Richard Lewontin, Niles Eldredge, and Stephen Jay Gould all come out of the zoological tradition, which suggests to me that ... they deal with a data set some three billion years out of date,' Margulis says. Animals, she points out, 'are very tardy on the evolutionary scene, and they give us little real insight into the major sources of evolution's creativity'.

Margulis has won few friends. In 2009, in a move that must have lost her most of her remaining admirers, she pulled the prestigious US National Academy of Sciences through the dirt and forced it to change some of its time-honoured – though extremely unscientific (one might even say anarchic) – practices. Thanks to an outcry raised by Margulis's 'abuse' of the system, members of the National Academy can no longer sidestep proper academic criticism and push a research paper, unresisted, through the peer review process. Some might say that, in the case of the National Academy vs marine biologist Donald Williamson, Margulis has done science a favour.

Donald Williamson is an Eeyore among scientists. In his own words, he is 'from a short-lived family, and on a straight-line course for posthumous recognition'. He doesn't have much luck, either. Shortly after he used these words to tell Lynn Margulis what he thought of his prospects, he slipped and fell while collecting specimens. Williamson, who is eighty-nine years old as I write, is now confined to a wheelchair. Whether he's right about the posthumous recognition, only time will tell. But it certainly doesn't look as though the recognition will arrive anytime soon. In 2009, one of Williamson's scientific peers dismissed his ideas as 'the most stupid thing that has ever been proposed'. Another suggested that a paper Williamson had submitted for publication in the *Proceedings of the National Academy of Sciences* would be better suited for the *National Enquirer*.

In one respect, though, Williamson is lucky. If people ever do the experiments he has suggested, and find that his 'astonishing and unfounded' ideas (another critic) are right, no one will be able to deny his claim to priority. The stink Lynn Margulis kicked up in her promotion of Williamson's work has made sure of that. In fact, it is hard to believe that Margulis didn't do this deliberately. As a member of the National Academy of Sciences, she was able to fast-track papers she likes through the peer review process. As long as she could find two reviewers who also liked the paper, the NAS pretty much guaranteed that they would publish it in the *Proceedings*.

That's how Williamson managed to publish his notion that the bodies of butterflies and their caterpillars have come down different evolutionary pathways. His claim is that the stark and mysterious differences between butterflies and caterpillars – and many other creatures that go through a larval stage before entering adulthood in a completely different form – are the result of a hybridisation in their distant evolutionary past. Sometime, long ago, a female creature's eggs were accidentally fertilised by sperm from another species. The result was the development of a species with two very different phases of life. In one phase the genome of one of the ancestral species controls what the body looks like. Then, at some trigger moment, the other species' genome takes over.

At first glance, it's a compelling idea. Take the starfish that biologists know as *Luidia sarsi*, for example. Luidia, like most starfish, starts life as a tiny star within a larva. The larva grows, and the starfish slowly moves to its outside. Where most starfish then emerge as adults, while their larval partners die and decompose, the Luidia splits into two perfectly happy organisms – a starfish and a larva – that separate and live independent lives. It is something like the successful separation of conjoined twins.

For Williamson, this is possible only if there are two genomes involved – and thus two very well-separated ancestors that really should have had nothing further to do with each other. No wonder Margulis loves the idea: it is a further desecration of Darwin's neatly branching tree of life.

Other biologists are much less keen, which is why Margulis struggled to find two positive reviewers. In August 2009 she told

Scientific American that she had gathered '6 or 7' opinions before she had what she needed to push the paper through. Unfortunately for her – though it is not beyond the realms of possibility that it was all part of Margulis's plan – Randy Schekman, the editor-in-chief of the *Proceedings of the National Academy of Sciences*, read that interview. Within days, all hell had broken loose.

Schekman wrote to her, asking what she thought she was doing. The rules say you can't cherry-pick: you have to submit all the reviews you get. Williamson's paper had already been published online (to near-universal derision); now Schekman was hesitant about whether it should appear in print. He also suspended other papers that had Margulis's endorsement.

Margulis threatened litigation. She also revealed her mischief-making: three of those 'negative' reviews never happened because the recipients of the paper were too busy or considered the subject outside their area of expertise. Then she added that she had asked a couple of amateur (but competent) naturalists their opinion too. 'My modus operandi is to ask competent people, whether or not they have a PhD,' she said. All in all, it looks as if she played them for suckers in the cause of getting an outrageous idea heard and discussed. Anarchy.

The *Proceedings of the National Academy of Sciences* no longer has a fast track for submissions. Though there was no announcement, the journal says it was going to stop the process anyway. The Margulis–Williamson affair must have been quite a catalyst. But there remains no guarantee that Donald Williamson will ever be vindicated, posthumously or otherwise. And not because of the Machiavellian manoeuvrings of scientists: he might be just plain wrong.

Lynn Margulis was right about endosymbiosis. However, her

current obsession, that HIV does not cause AIDS, is widely considered to be absolutely and categorically wrong. The Cornell physicist Thomas Gold was right when he speculated that pulsars were rotating neutron stars, but hopelessly wrong in his speculations about the origins of the universe. Cambridge University's Brian Josephson won a Nobel Prize for his insight into the properties of superconductors; his current ravings about the plausibility of extrasensory perception seem less well thought through. Being right once doesn't make you right the next time, not in scientists' eyes. The only people who are automatically 'right' are those who have somehow risen to the top and then clung on to establish a dominant hold on their field, kicking at those beneath them who attempt to climb up. Everyone knows that science is meant to be a meritocracy, but science is also human, and no one likes to give up a hard-won throne.

Here is a typical testimony from the writer and research biologist Jenny Rohn:

> After the very first talk I ever gave at an international symposium, one of the field's worthies rose to his feet in the hushed auditorium and proclaimed, with a scathing sneer, that my theory was completely misguided. I was too shocked to make the reasoned rebuttal that I could easily manage today, and too innocent to realise that the man's chief objection stemmed from the threat that my (ultimately true) findings cast on his own work.

Rohn adds that she has since seen 'many colleagues skewered on the podium in their turn'. Though she puts a positive spin on the experience – she adds that it 'is all part of the process of polishing truths out of rough ore' – there is no doubt that there is a downside to the humanity of science.

In every area there are 'worthies' who dominate discussion and shape the outcome of the peer review process. As Carl Lindegren has pointed out:

> One likes to think of science as divorced from personalities because one seeks the guidance of a principle rather than a person. Thus, the individual scientist experiences a feeling of freedom since he has the impression he lives in a community in which the law and not the man is the ultimate arbiter. This truly democratic practice has led to the fallaciously democratic practice of determining the validity of a scientific view by finding out how many other scientists agree with it. Voting in this context is so much influenced by past training and indoctrination that it tends to reject the new and to reaffirm the old.

So, having celebrated the victories of tenacious scientists, perhaps it is time we took a look at the darker side of the battle for scientific supremacy – the struggles of the outsiders who ultimately made their mark, but were never invited in. Sometimes, the secret anarchists can be immeasurably cruel.

7

DEFENDING THE THRONE

Machiavelli would be proud

High in the Sierra de Guadarrama mountains in central Spain, Robert Jordan is preparing to blow up a bridge. The detonation will help the Republican army to relieve the siege of Madrid, possibly the most horrific episode in the Spanish Civil War. For years now, the Nationalists have had the Republican-held city cut off. They have bombed it from the air, inflicting mayhem, injury and death on its hundreds of thousands of men, women and children. The survivors now lack shelter and food, and are so dispirited that they are contemplating surrender.

The need is pressing, but Jordan is wavering. To get the job done he may have to murder a hero: the once-great resistance fighter Pablo, who is no longer a help but a hindrance to the struggle. As if that weren't hard enough, Jordan knows that he is unlikely to get out alive himself. Blowing up the bridge will provoke a firefight. He will almost certainly die in the aftermath.

Should he do it? There are plenty of good reasons not to. After years in the wilderness, he has just found love. And this is not his fight: he is an outsider, an American brought in to support the struggle against the Fascists. His last act will be only a tiny contribution to this vast, heaving, bloody mess in a country thousands of miles from his home. Is it really worth losing your comforts, your hopes and dreams, over what you could justifiably call someone else's cause? This is the question Ernest Hemingway poses in his moving tragedy *For Whom the Bell Tolls*. It is a question that also faces the anarchists of science. What price discovery?

Science is civil war without the bloodshed. There are sieges, and there are bridges to be blown. There are people who must be removed: those who used to be heroes but are now complacent and ineffective must be forced aside for the good of the cause. But, like Pablo, some of this old guard still have arms and ammunition, and will fight to the very end. Many an anarchist lost their life fighting for the future of Spain. And many scientific anarchists know what it is to lose everything in the pursuit of discovery.

Take Chandrasekhar Subrahmanyan. He, like Robert Jordan, crossed continents to play his part in a battle. Yet, also like Jordan, much of his energy was taken up in dealing with a troublesome, belligerent but powerful rival. In Chandra's case, the establishment figure was the astronomer Arthur Eddington.

Eddington is generally regarded as one of the greatest British astronomers. We have already seen how he helped to make Einstein's name by confirming the predictions of general relativity – using some rather anarchic means. In 1935, though, his anarchy reached its peak. At the time, Eddington was Plumian Professor of Astronomy at Cambridge. He had received numerous prestigious awards: the Royal Society's Royal Medal, the Royal Astronomical

Society's Gold Medal and the Astronomical Society of the Pacific's Bruce Medal for outstanding lifetime contributions to astronomy. Everyone knew him as an exceptional astronomer, an unparalleled eye on the stars. When it came to Chandra, however, Eddington had a blind spot.

Chandra arrived in England from India in 1930. He brought with him a startling new scientific insight that had come to him in a flash of inspiration while his boat crossed the Arabian Sea. Sat on a deckchair, it took Chandra little more than ten minutes to perform the calculation that confirmed his suspicion: at the end of their life, the heaviest stars would collapse in on themselves ad infinitum. They would create rips in the very fabric of space and time, rips that we now know as black holes. The discovery earned Chandra a Nobel Prize – but not until 1983. Arthur Eddington may have died in 1944, but his public excoriation of Chandra's work had a long-lasting effect.

Hemingway describes the Spanish winter sky as 'hard and sharp with stars'. To our eyes, the stars do appear sharp: pinpricks of heavenly light piercing the dark curtain of the sky. It is difficult to conceive of the stars as the astrophysicist sees them: huge balls of gas, mostly hydrogen, that have been burning for billions of years.

Even more difficult to grasp is the fact that the stars' colossal energy output is derived ultimately from nothing more exotic than gravity. Once gravity has pulled enough hydrogen molecules together in one place, the ever-increasing proximity of the atoms to one another only increases gravity's influence. In a sphere of hydrogen hundreds of thousands of kilometres across, the atoms at the centre feel the weight of thousands of trillions of tonnes pressing in on them. The result is that these atoms fuse together, releasing vast amounts of energy.

Nuclear fusion is an astonishing phenomenon: alchemy in its purest form. In the core, or nucleus, of a hydrogen atom sits a proton, a lone positive charge. Force two hydrogen atoms together, and their positively charged protons will repel each other. But overcome this repulsion with enough pressure, and one of the protons will convert itself into a neutron, and it and the proton join to form a hydrogen isotope called deuterium. Combination with another neutron creates tritium, and when two tritium isotopes combine they fuse to form a helium nucleus. Heavier elements are formed in a similar way.

Every fusion event releases energy – lots of energy. After ignition, the ball of burning gas around the star's core will reach temperatures of millions of degrees. Stars can maintain this energy output for billions of years, but it can't last for ever. While the star is burning, the energy release creates an outward pressure that counters the crushing effect of gravity and holds up its outer layers. When all the fuel runs out, though, gravity takes control again, and the star collapses. All the atoms in the dying star are pulled towards the centre, and the closer they get, the stronger the pull. When a star has a large mass, something like twice the mass of our Sun, that ever-increasing pull will carry on until the star vanishes to nothing. This was Chandra's startling discovery: the star disappears from the universe. All that remains is the mysterious structure we know as a black hole.

Black holes have become a staple of science and of science fiction. Though we can't see them directly – because nothing, not even light, escapes their gravitational pull – NASA's telescopes have recorded their voracious nature in the X-rays emitted by particles spiralling inwards on their way to oblivion at the singularity, the heart of the black hole. The instrument that has done most to illuminate this process is the Chandra X-ray Observatory, a spaceborne telescope named in Chandra's honour and launched on the

Columbia space shuttle in 1999, four years after his death.

It is an honour that, according to his wife Lalitha, would not have impressed him. In his bittersweet book *Empire of the Stars*, Arthur Miller tells Chandra's story. Towards the end, he recalls meeting Lalitha after Chandra's death, and asking whether NASA's act would have meant something to her husband. He would have dismissed it, she said, with a casual 'So what?' Chandra had spent decades wanting to be part of the establishment, and the door had remained firmly closed. And it had all gone wrong from the start.

The grand unveiling of Chandra's discovery that dying stars could vanish from the cosmos was set to take place at a meeting of the Royal Astronomical Society on 11 January 1935. In the preceding months, Eddington had arranged for Chandra to have everything he needed for his work, including expensive calculating equipment. For weeks beforehand, Eddington had been visiting Chandra in his rooms at Cambridge, asking questions about the fate of the larger stars, how Chandra arrived at his conclusions, and what – exactly – he would present at the meeting. Chandra assumed that great importance was being attached to his presentation. And then, on 10 January, the secretary of the Society took Chandra aside and revealed that Eddington had put himself on the programme directly after Chandra's talk. Eddington's talk was to be called 'Relativistic Degeneracy', indicating that it had to do with Chandra's theory. That was the moment when Chandra suspected that he had been naive about Eddington's attentions. The events of the following day proved his suspicions to be well founded: Eddington's visits had been motivated, in Miller's words, by 'sheer mean-spirited duplicity'. Eddington stood up after Chandra's talk and told the assembly, 'The paper which has just been presented is all wrong.'

Chandra sat, astonished, while Eddington tore into his work. Eddington couldn't fault the mathematics, and he didn't bother to try: he simply ridiculed the basis of the idea that a star could disappear. His critique of Chandra's paper contained quips made at Chandra's expense. Eddington dismissed the work as 'stellar buffoonery'. He told the assembly, 'I think there should be a law of Nature to prevent a star behaving in this absurd way.' Chandra was humiliated; he later recalled laughter breaking out at many points during Eddington's talk. Eddington, he said, 'made me look like a fool'. At the end of the meeting, Chandra's peers offered their com-miserations. 'The other astronomers were certain that my work was wrong because Eddington had said so,' Chandra recalled.

Later in the year, at a conference in Paris, the humiliation was repeated. 'Eddington gave an hour's talk, criticising my work extensively and making it into a joke,' Chandra said. He asked the American astronomer Henry Norris Russell, who was presiding over the session, for a chance to reply. 'I'd rather you didn't,' was Russell's response. It looks like a spineless caving-in, but Russell was simply being loyal to his mentor.

In October 1977, the historian Spencer Weart sat down with Chandra and recorded an extensive interview. They discussed the history of astronomy and the way Chandra got his ideas. Most interesting, though, was Chandra's appraisal of Eddington, and of Eddington's status. 'Oh, his position in astronomy was dominant,' Chandra told Weart. 'I don't think there was any doubt in any-body's mind that Eddington was always right.' Chandra's ordeal had made a small dent in Eddington's reputation, though: some of their colleagues acknowledged – in confidence – that, on this occasion, Eddington was wrong. For weeks after the talk at the Royal Astronomical Society, Chandra's colleagues would con-sole him in private with a mumbled apology that conveyed their faith in the black hole idea. But they were covert confessions. 'Of course all these people who supported me never came out publicly,' Chandra said. 'It was all private.'

The whole experience was to have a lasting effect: six years before he received notice of his Nobel Prize, Chandra confessed his lowered expectations to Weart. Scientists, he said, make a memorial for themselves through their work. Some make discoveries, but others take on a more modest role: to gather information and material that may prove helpful to others. 'I have chosen the latter approach,' Chandra told Weart. 'All, I think, as a consequence of my first shattering experience in Cambridge.'

What was behind Eddington's vicious denial of Chandra's work? The black hole theory was certainly problematic for Eddington; Chandra's calculations on the fate of white dwarf stars produced numbers that undermined Eddington's ongoing efforts to create a grand unified theory that described everything from atoms to stars. But that alone was not enough to provoke the kinds of merciless attack Eddington meted upon Chandra.

Miller hints at a homoerotic component to Eddington's behaviour; rumours abounded that Eddington was homosexual, and thus psychologically compromised in a society where he would be ostracised – and possibly prosecuted – if this came out. If he was already under huge emotional strain, Miller suggests, perhaps Chandra's derailing of Eddington's attempts at producing a grand unified theory would have been more than he could bear. It's an engaging idea, but, as Miller admits, Eddington didn't have *that* much to lose, career-wise, by accepting Chandra's idea. He had even stumbled across the idea that a heavy star would eventually disappear to 'nowhere' himself, and had dismissed it as simply nonsensical.

The real reason for Eddington's venomous hostility to Chandra's work is probably more prosaic. Eddington presided over the British astronomy establishment, and saw Chandra as an arriviste from the colonies. It seems to be a simple case of insider versus outsider; racism, in effect – the privileged Englishman blocking the dark-skinned Indian from gaining admission to the club. Chandra was, most likely, a victim of the Raj mentality. The British Empire was still a major force in the world when Chandra arrived from Madras. With his dark skin and idiosyncratic English, no one at Cambridge would have looked upon him as 'one of them'. He encountered overt racism at Cambridge, and no British university offered him a permanent position, even though many such positions were open.

In the face of such bleak prospects, Chandra made a difficult decision. Rather than fight, he ran. Eddington and the rest of the English astrophysicists successfully drove him away from their territory. He went to work in the United States, and in a different field of astronomy: he became what Miller terms a 'reluctant astrophysicist'. Theoretical physics was an avenue that Eddington had closed off for him. As he said to Lalitha, 'it is because of Eddington that I became the sort of scientist changing field periodically from one to another. I had to change my field after the controversy'.

The Nobel Prize announcement came nearly half a century later. 'It's about time,' Lalitha declared, but Chandra was ambivalent about the award, and declined an invitation to a party in his honour.

For most of us, goings-on in outer space are not of huge interest. If asked to describe what is out there, just beyond the edge of the Earth's atmosphere, we might come up with descriptions of calm emptiness, blackness and silence. Small wonder, then, that scientists pretty much ignored the Earth's immediate environment for centuries. In the days before Sputnik and the space race, it was considered dull and empty. Meteors would whizz through the heavens, and the spectacular northern lights would brighten the Arctic skies every now and then. But there was little else for scientists to get excited about. And then along came the Swedish physicist Hannes Alfvén.

Alfvén lived a quiet domestic life. His wife, Kirsten, died after they had been happily married for sixty-seven years and had raised five children. He loved to study Eastern philosophy and, towards the end of his life, to wait on the beach at sunset watching for the green flash of refracted light as the Sun disappeared beneath the horizon.

Perhaps it was this quiet domesticity that allowed Alfvén to be such a revolutionary in his professional life. He continually broke new ground after invading fields of research in which he had no formal qualifications. He summarily dismissed the views of experts, holding no truck with 'received wisdom'. He never waited to be proved right before moving on to wreak havoc elsewhere. He received little acknowledgement for his contributions; even the physicists who used his work had no idea where it came from. That is largely because of his status as an outsider in the areas in which his contributions were most valuable; he was denied the privileged status his remarkable achievements should have brought him. Alfvén's Nobel Prize was for work he had carried out in the 1930s, but it was not awarded until 1970. It is surely no coincidence that this was just a few months after the death of one Sydney Chapman.

Shortly after the announcement that Alfvén was to be awarded a Nobel, the physicist Alex Dessler wrote a telling piece in the journal *Science*. 'For much of Alfvén's career, his ideas were dismissed or treated with condescension; he was often forced to publish his papers in obscure journals; and he was continually disputed by the most renowned senior scientist working in the field of space physics,' Dessler said. That 'renowned senior scientist' was Sydney

Chapman. If the story of space physics were to be made into a superhero strip cartoon, Chapman would be portrayed as Alfvén's nemesis. A British mathematician who applied his expertise to the physics of space, Chapman was, like Eddington, a lynchpin of the establishment. And he was not afraid to use his status to keep Alfvén in his place.

Chapman was a fellow of the Royal Society, a member of the US National Academy of Sciences, a scientific advisor at the Geophysical Institute of Alaska, and an honorary fellow of Queen's College, Oxford, Trinity College, Cambridge, and Imperial College, London. 'It would be difficult to overestimate the great influence which Chapman exerted on the scientific world at large,' the writer of his obituary in the London *Times* noted, adding that 'Chapman's mild manner veiled a strong will and great determination.' What is not mentioned in that obituary is his anarchic behaviour towards Hannes Alfvén. Science is meant to be a level playing field, with each idea standing on its own merit, regardless of its source; its status and acceptance should be subject only to the test of experiment. That was certainly not Alfvén's experience.

Dessler followed up his statement about Alfvén with a confession: he had been one of Chapman's unthinking disciples. It was only when he was persuaded to – by one Subrahmanyan Chandrasekhar – that he was 'shamed' into looking at Alfvén's work objectively. In that moment, Dessler had what can only be described as a road to Damascus experience. 'My degree of shock and surprise in finding that Alfvén was right and his critics wrong can hardly be described,' he wrote. Alfvén's work had been 'drowned' by Chapman, he observed. 'How could this have happened? After all, do we not believe in the objectivity of science?'

Those in the know never did believe in it. The great German physicist Max Planck once noted that 'A new scientific truth does not triumph by convincing its opponents and making them see the light, but rather because its opponents eventually die.' If the fact that Alfvén's Nobel Prize came shortly after Chapman's death is no coincidence, it is surely no accident that Chapman's own ascendancy began at the untimely end of another scientist's life.

On 15 June 1917, the Norwegian physicist Kristian Birkeland was found dead in a Tokyo hotel room. He had swallowed twenty times the recommended dose of his prescribed sleeping tablets, the barbiturate Veronal. It may have been suicide, but it seems more likely to have been the result of a psychotic episode induced by long-term use of the drug; seeing its effects on their patients, many doctors had long ceased prescribing it. Whatever the truth, it was a sad and ignominious end to the career of a man who had been crowned the King of Space.

Through a series of laboratory experiments, polar expeditions and mathematical calculations, Birkeland had worked out that the aurora borealis – the northern lights – were the result of an interaction between the Earth's magnetic field and space-borne electric currents. 'It seems to be a natural consequence of our points of view to assume that the whole of space is filled with electrons and flying electric ions of all kinds,' he wrote in 1913. 'It does not seem unreasonable therefore to think that the greater part of the material masses in the universe is found, not in the solar systems or nebulae, but in "empty" space.'

Those electrons and flying electric ions, Birkeland surmised, must come from the Sun, but his theory could be tested only by making measurements in space. At a time when the aeroplane was just a decade old, that was far beyond science's capabilities, and it was not until 1963 that Birkeland was proved right. And that is partly because, in the intervening years, Sydney Chapman seized the throne of space physics. Birkeland's expeditions had established that aurorae occur as a result of electrons flowing down through the atmosphere along the lines of the Earth's magnetic field. Chapman, though, had other ideas. He had developed a theory in which electrons moved only through the ionosphere, one of the outer layers of the Earth's atmosphere.

In Chapman's scheme there was no downward movement, so it made no headway in explaining why the aurorae are visible from the Earth's surface. However, his mathematics was beautifully constructed and reproducible by anyone with a basic skill in maths. This was the root of his appeal: his approach to space physics was to simplify the situation until solvable equations emerged. Mathematically challenged space physicists loved it and took it on board. And that is why, when Hannes Alfvén offered a complex but rigorous mathematical support for Birkeland's idea of currents that travelled down towards the Earth's surface from the ionosphere, it was rejected out of hand. The editors at *Terrestrial Magnetism and Atmospheric Electricity*, the leading American journal of space physics, were at least candid about the reason for rejection. Alfvén's calculations, they said, could not be right because they did not agree with Chapman's calculations. That Chapman might have been wrong did not occur to them. Or if it did, that wasn't a road they were prepared to go down.

Chapman's influence was astonishingly far-reaching. 'I have no trouble publishing in Soviet astrophysical journals, but my work is unacceptable to the American astrophysical journals,' Alfvén once said. In the end, he managed to publish what is now the foundation of our modern understanding of what goes on above our heads in a Swedish-language journal called *Kungliga Svenska Vetenskapsakademiens Handlingar*. It's unlikely that you – or most physicists, for that matter – have ever heard of it. And yet it contains the original paper on what might go wrong when an angry Sun spits a well-aimed flare towards the Earth.

In January 2009, the US National Academy of Sciences issued a research report on the dangers posed by the Sun. To be precise, the report was issued by the Committee on the Societal and Economic Impacts of Severe Space Weather Events. But this was no minority interest issue: the research was carried out by some senior figures at US universities, and funded by NASA. The committee concluded that there was a chance that, in the next few years, the United States could be brought to its knees by solar activity. If the Sun flings out one fortuitously aimed gob of material, it could cause a geomagnetic storm powerful enough to leave half a continent without electrical power.

In our modern, technological society, electricity supply is vital. When particles spat out by the Sun in a violent 'coronal mass ejection' interact with the Earth's magnetic field, the result can be chaos. The interaction can induce enormous currents capable of melting wires in transformers in power transmission networks, causing breakdowns that can take a year or more to repair. Committee members familiar with the workings of the electrical grid and the likely impact of its failure warned that the consequences of such long-term failure would be extraordinary. The supply of drinking water, fuel, heating, food on supermarket shelves and vital medicines ultimately relies on electrical power. The report estimated the cost of damage, loss of life and loss of economic output to be as high as $2 trillion in the first year. To put that in context, Hurricane Katrina's financial impact in 2005 was estimated to be somewhere around $100 billion. It is questionable, the researchers concluded, whether the United States could ever recover from such a blow.

There has been little reaction to the warning. Perhaps that is appropriate. Although there are measures we could take – all technologically advanced nations, not just the United States, are at risk – the chance of this total wipeout happening are low enough to make it a threat we might get away with ignoring. Examine the history of this field of research, however, and it is hard not to conclude that the blind eye turned to the impact of a geomagnetic storm is not a measured response. It is the consequence of a long history of snubs to the physicists who work in this area.

That National Academy of Science's report on a possible space-borne apocalypse stems from Alfvén's obscure but ground-breaking paper. We now know Alfvén's roiling streams of charged particles as plasma. The Sun itself is composed of plasma, and our knowledge of plasma physics forms the basis of our understanding of the Sun and its interaction with our atmosphere. This whole field, known informally as space weather, is now of vital interest. Thousands of artificial satellites now orbit our planet and perform a variety of tasks essential in modern life – TV broadcasting, navigation, military surveillance, telecommunications, weather and climate forecasting, for example – and they are extremely vulnerable to space weather: a heavy shower of plasma can fry a satellite's electronics.

But Alfvén's legacy is not just about space weather. For a start, plasmas are central to much of physics. Particle accelerators such as CERN's Large Hadron Collider create and analyse plasmas. Future technologies, such as fusion reactors that aim to release nuclear energy without an explosion, are all about learning to control and manipulate plasma. Alfvén made contributions in other fields, too. In 2010, President Obama gave NASA a new objective: to travel to an asteroid in order to probe its structure, which might give clues to the history of the solar system. Coming as it did from the mouth of the President, it sounded like a new and exciting idea. But go back to Alfvén's Nobel lecture, given on 11 December 1970, and you'll find the same suggestion.

Hannes Alfvén has been called a heretic, a dissident, an iconoclast and an enigma. He is also a heroic resistance fighter. With his defeat of Sydney Chapman and in establishing so many branches of physics, Alfvén slew his Pablo and blew the bridge. And he kept on fighting: he dedicated much of his later life to campaigning for nuclear disarmament, and was vociferous in his dissent against the Swedish Government's take-up of nuclear power.

An anarchic spirit, you see, is not necessarily a bad thing. The story of the Spanish Civil War is largely a tale of atrocity, oppression and massacre, but there were also moments that created a light in the darkness. One such light is still shining, in fact: the anarchic spirit of FC Barcelona.

At the outbreak of the Spanish Civil War, the people of Barcelona took control of the city's street railways. Anarchy ensued: not anarchy as disorder or chaos, but anarchy as removal of the ruling classes. The workers' union took over the railways and dismissed the directors, who had been paid eighteen times what the average railway worker earned. With the directors gone, the wages of the lowest earners went up 50 per cent. What's more, the unions were able to radically improve the efficiency of the system. Fares were reduced, and certain members of the population – schoolchildren, invalids, wounded soldiers, those who had suffered injuries at work – were allowed to travel for free. What started on the railways soon made its way to the ports, the utility companies, the clothing industries – even the hairdressers. Catalonia became an anarchist state. In many ways, it still is.

Nowhere is the Catalan anarchy more apparent than in the ethos of its pride and joy. FC Barcelona has one of the best football teams in the world, and is owned and operated entirely by its supporters, who seek to be a force for good. Its shirts, for example, carry the logo of the United Nations Children's Fund, UNICEF – and it pays for the privilege. Every year, the club donates 0.7 per cent of its income to UNICEF. This is no arbitrary figure: it is the proportion of gross domestic product that the UN wants to see rich countries contributing to international aid. The anarchy of science can be similarly world-enhancing, even as it flies in the face of capitalism and bureaucracy. Just ask Stanford Ovshinsky, a man who has been hailed as America's greatest inventor since Thomas Edison.

In November 1968, one of Ovshinsky's inventions, known as the threshold switch, made the front page of the *New York Times*. It would, the *Times* said, lead to 'small, general-purpose desktop computers for use in homes, schools and offices', and 'a flat, tube-less television set that can be hung on the wall like a picture'. But no one would invest. Why? Because Ovshinsky wasn't a bona fide scientist – he didn't even have a college degree. No one in American science trusted this outsider.

Ovshinsky's parents were immigrants from Eastern Europe and raised their son in a backwater in Ohio, where Ovshinsky Senior made a living by going through the streets collecting scrap in a cart. After high school, the young Stanford Ovshinsky became a trainee machinist in a local factory. But he didn't abandon his education: each week he took home an armful of books from the Akron Public Library. While he was at work he thought about ways to improve the efficiency of the factory's machines – and he then built a prototype lathe which, he claimed, would do just that. Everybody laughed until he turned it on and showed how well it worked.

Shortly afterwards, he set up his own machine tools company, then sold it to the New Britain Machine Company. A few years later, Ovshinsky's lathe was saving the lives of US soldiers in Korea. The army had run out of artillery shells; only by employing Ovshinsky's high-efficiency lathe could it meet demand and keep the North Koreans at bay. Ovshinsky, meanwhile, had moved on.

At the time, all focus in electronics research was on the rigid, crystalline properties of semiconductor materials such as silicon.

Semiconductors, the basis of transistors, are largely crystalline: their atoms are arranged with rigid, military precision. Metals are even more ordered, which is what allows them to be such efficient conductors of heat and electricity. Electrons can move with ease through the regular, crystalline atomic lattice that forms the structure of metals; they are not continually banging up against some unexpected atom. The semiconductor revolution took off only when researchers at Bell Labs learned to grow large enough crystals of the 'semi-metal' germanium, free enough of electron-impeding defects for the crystals to flip between conducting and insulating behaviour at the flick of a switch. In the 1960s, silicon technology, and the transistor in particular, was turning America into an electronic superpower. Ovshinsky watched this development with interest, and then, characteristically, turned it on its head.

Perhaps it was when he used to watch the scrap metal dealers melting down his father's rigid metal finds into liquid, or the time he spent in the factories, machining metal bars, cutting away at the metal's crystalline structure. Maybe it was simply the frequency with which the metal blades on his lathes became blunt. Whatever the cause, Ovshinsky didn't have the reverence for crystals that characterised the mindset of the scientific establishment. He wondered whether disordered, messy materials might prove just as useful.

The materials that interested Ovshinsky are properly called 'amorphous': without defined shape or structure. The glass in your window is amorphous: the silicon dioxide molecules in the panes that let light into your house are arranged in no particular way. If you were to look at them under an electron microscope, you would see a jumble of molecules: there is no order beyond the tetrahedral arrangement of the silicon and oxygen atoms that make up each molecule. The idea that amorphous materials might prove useful was bolstered by Ovshinsky's new interest. His reading had led him to a fascination with the brain. If the messy, disordered structure of brain cells produced the powerful organic computer inside our skulls, he reasoned, then perhaps materials without an ordered structure could also have electrical properties as interesting as those being found by researchers in the silicon industry.

Ovshinsky knew that the transistor works by controlling a crystal's conductivity. If you apply energy in the form of an electric field, you can switch the crystal from blocking the flow of electric current to conducting it. Perhaps, he thought, energy – whether light, electricity or even heat – could do similar things to an amorphous solid. Perhaps it would jolt the atoms into a more useful arrangement, and give the solid some interesting properties. Perhaps it would even be reversible. His early experience with the development of the lathe had taught Ovshinsky that crazy ideas need more than an enthusiastic advocate: they need a prototype. And so he set about trying to create his own amorphous version of the transistor.

Eventually, it worked. Ovshinsky's threshold switch was a thin film of amorphous material sandwiched between two metal conductors. When a large enough voltage was applied across these conductors, the amorphous material would become a crystal. That changed it from being an insulator to being a conductor, just like that. When the voltage was reduced, it would switch back.

As if that wasn't revolutionary enough, Ovshinsky created another material in which the switch between conducting and insulating states didn't need a continuous flow of electrical power. You hit the material with a burst of electricity, and it went from one state to the other and would stay in whatever state you left it until you applied another voltage. The technology is a kind of non-volatile memory, the kind used in the memory sticks that you plug into your USB drive to transfer stored data. Ovshinsky's patent has now been licensed by the chip-maker Intel, among others. In the United States of the 1960s, however, no one who mattered put any credence in Ovshinsky's claims. Getting amorphous materials to switch between conducting and insulating states was thought to be impossible, so no one believed Ovshinsky when he said he had got it to work.

A February 1970 article in the magazine *Science & Mechanics* illustrates the reaction of the scientific establishment. It was called 'The Ovshinsky Invention', and examined his claim that if you applied a voltage to a particular type of glass, it would conduct electricity. If this was true, it was a discovery that would threaten the newly established dominance of the silicon transistor. The article's subtitle made clear the mood of the time: 'Is it greater than the transistor, or is this self-taught engineer a fraud as the big companies claim?'

It was this kind of hostility and scepticism that sent Ovshinsky, who was now close to bankruptcy, to Japan. His early inventions were licensed by Canon, Sony, Sharp and Matsushita – companies that became the superpowers of the Japanese electronics industry. Thanks to Japanese investment, Ovshinsky's inventions now occupy a key place in the technological landscape. The flat LCD display in your TV and your computer monitor, a screen that really can be 'hung on the wall like a picture', is a development of his amorphous silicon technology. You might also be a fan of the rewritable CD and DVD, an invention for which Ovshinsky received the key patent in 1970.

When he invented the nickel metal hydride battery, also based on amorphous materials, that too got massive take-up from Japanese manufacturers who ended up dominating the battery market: 'NiMH' is now written on the side of billions of batteries sold every year. Ovshinsky's idea of using a cheap form of amorphous silicon to build a solar panel was also welcomed in Japan

– particularly by Sharp, which was selling solar-powered calculators by the container ship-load.

Small wonder that Ovshinsky has been referred to as 'Japan's American genius'. But despite the deals he made there, he never became wealthy. Perhaps that is why *Forbes* magazine has called him 'the inventor who can create anything but profits'. It's a fair description. In 2008, for example, Intel teamed up with STMicroelectronics to create a new company called Numonyx. Two years later, Numonyx was sold for $1.3 billion. What made it so valuable? This company manufactures hard disk and flash memory units for cameras, phones and MP3 players. The new technology, widely seen as the successor to flash memory, is properly called 'ovonic unified memory'. It is based on amorphous silicon. The term 'ovonic' is a contraction of Ovshinsky Electronics. After being rejected by the scientific establishment for being an untrained outsider, Stanford Ovshinsky has stamped his name on billion-dollar industries. And yet he has never made money.

Ovshinsky is an old man now. His rare interviews in the press always mention his shock of snow-white hair, sometimes described as a halo. The other thing that is always mentioned is the fact that Ovshinsky thinks like no one else in the business world. When he and his wife Iris set up Energy Conversion Devices in 1960, their aim was to use 'creative science to solve societal problems'. For nearly fifty years they did just that, developing solar cells that were cheap and easy to produce, innovating for the hydrogen economy, where cars run on hydrogen derived from water, and generally attempting to make the world a better place. Making money, it seems, was never on their agenda. In 2000, the Institute for Policy Studies analysed executive pay and found that CEOs were getting, on average, 500 times the salary of their average worker. Ovshinsky, on the other hand, was taking just five times the wage of those on the shop floor of his company. He was also still a member of their union.

The character trait behind this secret anarchy is the same one that earned Nevill Mott his 1977 Nobel Prize in Physics. He received the award for working out the electronic properties of certain crystals, but Mott felt he couldn't take the credit for the original idea: that was due to Stan Ovshinsky. 'A lot of my best ideas came from Stan,' Mott once told a friend. 'He just gave them away to me.' Mott was not the only beneficiary of Ovshinsky's generous nature. 'All of us in the field have had that experience,' says Stanford University's Arthur Bienenstock. Ovshinsky was not bitter about being left out of the Nobel Prize citation; on the contrary, he sent Mott the biggest bottle of champagne Mott had ever seen. 'I'll have to throw a party for fifty people,' Mott told a reporter from *New Scientist*. It was typical of Ovshinsky's generous, anarchic nature.

Not that everyone appreciates it. In the summer of 2007, the board of directors of Energy Conversion Devices and its investors decided that they'd had enough of not turning a profit and kicked Ovshinsky out. The tactic worked: the company's stock suddenly went through the roof. Less than a year later, ECD – now free to exploit Ovshinsky's insights without a social agenda – reported a profit for the first time in years. The truly astonishing thing is that Ovshinsky didn't even complain. He simply set himself up a new workplace right next to his Michigan home. His office at the Institute for Amorphous Studies is a replica of his office at ECD. The focus of the room is a huge wallchart of the periodic table of the elements, an exact copy of the one at ECD that seeded all Ovshinsky's great discoveries. 'I know what I want, I know what I'm going to do, and I use the periodic chart of atoms as if it's an engineering diagram,' Ovshinsky once said.

His creative use of science will go on as long as Ovshinsky has the power to think, but he will never become an insider. It's not just the lack of formal scientific training that has been an obstacle to Ovshinsky's acceptance. He has also made a rod for his own back by refusing – like Hannes Alfvén – to stick to one scientific discipline. He has published papers in neuroscience, cosmology, physics, chemistry, materials science, computer science and psychiatry. To Ovshinsky, the idea of separating science into disparate disciplines is unnatural and unprofitable.

Alfvén once explained why, in his opinion, the scientific consensus went the other way. 'Scientists tend to resist interdisciplinary inquiries into their own territory,' he said. 'In many instances, such parochialism is founded on the fear that intrusions from other disciplines would compete unfairly for limited financial resources and thus diminish their own opportunity for research.'

Sadly for science, this resistance to outsiders – the siege mentality – is usually rather successful. In the Gospel according to St Matthew, Jesus makes a telling statement about the economy of heaven: 'Whoever has will be given more, and he will have an abundance. Whoever does not have, even what he has will be taken from him.'

That's a vexing idea, and it goes against our deepest notion of fairness. Much more comfortable is Karl Marx's agenda for redistribution: 'From each according to his abilities, to each according to his needs.' However, as sociologist Robert Merton pointed out way back in 1966, although science ostensibly follows Marx, in reality it follows Jesus. The greater your scientific reputation, the more likely your papers are to receive rapid recognition. Once you've reached the top in science, it's actually quite hard to fall, even when you're busy treading on the fingers of those trying to assail your lofty tower. Merton calls this phenomenon of outsiders being kept on the outside – with much gnashing of teeth – the Matthew effect.

J.B.S. Haldane noticed it at work in his sphere. During the late 1950s he spent a few years as a professor at the Indian Statistical Institute in Calcutta, where one of his students, S.K. Roy, had carried out a Herculean task in improving the quality of various strains of rice. Haldane knew what would happen when he and Roy published the work together. 'Every effort will be made here to crab his work,' he wrote. 'He has not got a PhD or even a first-class MSc. So either the research is no good, or I did it.' According to Haldane, Roy deserved about 95 per cent of the credit. 'The other 5 per cent may be divided between the Indian Statistical Institute and myself,' he said. 'I deserve credit for letting him try what I thought was a rather ill-planned experiment, on the general principle that I am not omniscient.'

That may seem extraordinarily generous for a senior scientist, but then Haldane was a Marxist, and had been a fully paid-up member of the British Communist Party. Not that you have to espouse Marxism to be generous *and* a scientist. An earlier example of bucking the trend was the mathematician Isaac Barrow. In 1669, Barrow gave up the Lucasian Chair of Mathematics at Cambridge to make way for his student, Isaac Newton. It is interesting to note that Subrahmanyan Chandrasekhar dreamed of becoming the Lucasian Professor; sadly, the Machiavellian antics of Arthur Eddington made that impossible. Perhaps we shouldn't be surprised that Eddington, a devout Quaker, followed Jesus's philosophy to the letter.

Science is a fight to the intellectual death, but not between equal adversaries. It takes place in a gladiatorial arena where the challenger has to overcome not only the established champion, but also his or (more rarely) her supporters. And, whether in attack or defence, the fight is rarely clean.

We have seen the fraud that is 'normal misbehaviour'. We have seen how scientists conjure up an idea from somewhere no one else could have been, through drugs or mysticism or hallucinations or religious faith. We have seen that polished powers of persuasion, a silver tongue, can do wonders for the acceptance of your idea. Sometimes, though, you just have to be a pain. You face up to sheer, bloody-minded obstinacy with a fighting spirit. You don't give in to the belittling by your peers or even your superiors; you don't just give up on your 'hopeless' or 'misguided' idea. You find ways to beat the system. That's why science is not for the meek and mild. It is red in tooth and claw; its very ideas and breakthroughs are subject to the law of the survival of the fittest. Good scientists must strive to overthrow, undermine and destroy their colleagues' reputations. It's all neatly summed up in a quote attributed to the American playwright Gore Vidal. 'It is not enough to succeed,' he said. 'Others must fail.'

In some ways, Chandra, Alfvén and Ovshinsky did fail. They never managed to achieve the status of 'insider'. Though their scientific insights are now accepted and recognised, all three experienced the bitter taste of disappointment in their peers. Chandra remarked that, despite the Nobel Prize, he never considered himself part of the 'astronomical establishment'. Alfvén spoke with bitterness of his status as a 'dissident'; it was always, he said, a 'very unpleasant situation'. Ovshinsky is the least bitter of the three, perhaps because he always knew he would never get a scientist's ultimate recognition, the Nobel Prize. 'I'm not a part of their world,' he says.

The fact that the walls were too high for these scientists to climb is something the secret anarchists need to appreciate, because they are now faced with a task that requires those walls to be pulled down in a move towards unity and co-operation. As we will see in the final chapter, it is going to take a new and special kind of anarchy to get this job done.

8

IN THE LINE OF FIRE

Life on the barricades

From the outside, it seems an idyllic scene. It is the autumn of 2010, and the chestnut tree is just beginning to yellow against the red brick of Chicheley Hall, a magnificent Georgian country house set in 80 acres of beautiful gardens in rural Buckinghamshire. The Royal Society, the oldest scientific society in the world, has recently purchased the property for a cool £6 million and spent another £10 million converting it into a conference centre. The Kavli Royal Society International Centre will, they hope, offer scientists an atmosphere of relaxed creativity in which to work.

Inside the conference hall, however, the atmosphere is anything but relaxed. David Brin, a planetary scientist and science-fiction author, is fuming. His mouth is set in an ugly grimace, and from time to time his head shakes in exasperated disbelief. His eyes hardly lift from the table in front of him, an angry stare burning into the wood. It is clear that when the current speaker, Seth Shostak, Senior Astronomer at the alien-hunting SETI Institute, finishes, Brin is going to explode.

Shostak and Brin are taking part in a panel discussion about whether we should attempt to communicate with aliens. After years of listening – in vain – for alien signals, Shostak is keen that we send out signals from the Earth in a systematic effort to make ourselves known to the universe. Brin thinks that could prove suicidal. When his turn to speak comes, Brin turns on Shostak, all guns blazing. Shostak has displayed a 'stunning ignorance and an incredible lack of imagination', he says. Shostak, head slightly bowed, shows only a neatly cropped head of white hair and a wry smile. He has heard it all before.

As Brin's comments continue to pour contempt on Shostak's management of the search for extraterrestrial intelligence, it becomes clear that these two have history. Brin makes references to Shostak's 'arguments from ridicule' and his overseeing of 'Potemkin, stage-managed, party-driven Fox News-style meetings'.

This latter jibe is a reference to the times when the pair worked together. Brin was part of an effort by the International Institute of Aeronautics to draft a protocol about whether to broadcast messages into space in an attempt to reach out to alien civilisations. He eventually resigned from the committee in 2006 when Shostak and some other members changed some of the agreed wording, lopping off the proviso that international agreement should be reached before any such messaging took place.

'If they're silent,' Brin says (he means the aliens), 'then maybe they know something we don't know.' He is not scared of bogeymen, he says, he is properly cautious. And he is offended by the way he and people who share his sensibilities have been ridiculed:

'What we care about is the rudeness that's been going on, the failure of wisdom.'

As Brin concludes, the chair opens up the debate. Shostak is happy to get into an argument. If you think it's dangerous to broadcast, he says, you might as well shut down the search for alien signals too. The aliens, he says, 'will have reached the same conclusion'. Brin is still exercised. Back in 1990, he says, 'we were all one happy family here', a stark contrast to the 'mania of the last seven or eight years'. Shostak has had enough. 'You have provoked me,' he says, his voice rising in anger. 'No one was tyrannised.'

A Hungarian professor in the audience stands up and makes a plea. 'Can't we get back to the science?' he says. 'This looks like a TV show!' And he's right. It's like watching Jerry Springer host a science special. What's interesting – and rather funny – is what it takes to clear the air. About twenty minutes into the quarrel, an anthropologist, Kathryn Denning of York University in Toronto, stands up and asks a question about what level of broadcast signal would be detectable. 'I've been watching this debate for a number of years now,' she says. 'People with apparently equivalent credentials, and good brains with the ability to do math that I can't do, don't agree. Why not?'

There is a moment of stillness, as when a parent walks in on a fight between siblings. Then the astronomers close ranks. The agreement is pretty close, one says. No, says another, it's closer than that. A general hubbub rises in the room. There is no disagreement, they say. They are all friends again. Brin starts to talk about how good Shostak is on this topic and says that he listens 'very humbly and respectfully' whenever Shostak talks about SETI; Shostak has a 'wonderful paper' coming out on exactly this subject. The *volte face* is truly remarkable.

Nothing takes the fight out of scientists like the scrutiny of outsiders. They are secret anarchists, you see; open anarchy goes against the grain. Nonetheless, their concerns do occasionally run too deep to be contained, and then their anarchy is unleashed on an astonished world.

On 5 February 1987, one of the world's best-known scientists was arrested in Nevada. Carl Sagan had been trying to scale a fence and enter the area where the United States military puts its nuclear arsenal to the test.

The arrest was a direct consequence of Sagan's scientific studies. Four years earlier, he had attempted to pull together everything that was known about the aftermath of nuclear explosions. He applied this understanding to the scenario of all-out nuclear war and summarised the arguments in an article entitled 'The Nuclear Winter'. Any such conflict, he said, would most likely involve 'the explosion of 5,000 to 10,000 megatons – the detonation of tens of thousands of nuclear weapons that now sit quietly, inconspicuously, in missile silos, submarines and long-range bombers, faithful servants awaiting orders'. The result, Sagan concluded, would be the rapid death of around half the humans on the planet. Those who survived would have to live in near darkness for months as ash and dust filled and blackened the sky. Plants, unable to harvest enough light for photosynthesis, would cease to grow. Starvation, radiation sickness, looting and barbarous anarchy would be the inheritance of the survivors.

Sagan knew there was a wide margin of error in his calculations, but he presented all the possibilities; the 'tradition of conservatism' generally works well in science, he said, but is 'of more dubious applicability when the lives of billions of people are at stake'. So he published his findings. His analysis was greeted with anger by many atomic scientists, and acute lack of interest from government officials. Aware that he wasn't going to get far by using the standard scientific channels, Sagan joined a group of people who felt that the stakes were high enough to merit direct action.

At the time of his arrest, the United States was continuing with a programme of weapons testing, even though the Soviet Union had called a unilateral halt to such tests. Two years earlier, Mikhail Gorbachev had announced that the fortieth anniversary of the bombing of Hiroshima – 6 August 1985 – would mark the beginning of a Soviet moratorium on the testing of nuclear weapons. President Reagan declared the move nothing more than a propaganda exercise and refused to follow suit.

On that day in 1987, more than two thousand people gathered at the Nevada test site ahead of the year's first nuclear test. Sagan and 437 others were arrested and bussed to nearby Beatty, Nevada, where they were booked, charged with trespassing or resisting arrest (or both) and then released pending trial.

Carl Sagan's fight was against the misguided belief that scientists should not interfere with how their work is applied. He did not endorse the post-war rebranding of science and was not prepared to behave like one of the timid monks that Jacob Bronowski had identified as representing the new, craven spirit of science. It was not just in the sphere of nuclear proliferation: Sagan wanted the walls surrounding science to come down, and did his utmost to communicate the delight, the findings and the implications of science to the public. Sadly, it won him few friends among scientists.

Sagan was passed over for tenure at Harvard and denied membership of the US National Academy of Sciences after becoming famous for talking directly to the public through books, magazine articles and TV programmes. The atomic physics pioneer Edward Teller once spluttered to Sagan's biographer Keay Davidson that Sagan was 'a nobody' who 'never did anything worthwhile'. Yet when *Scientific American* columnist Michael Shermer took it upon himself to analyse the truth of this statement, he found that, in terms of peer-reviewed publications in journals, Sagan was among the greats. For lifetime publications he ranks alongside Jared Diamond, E.O. Wilson and Stephen Jay Gould. From 1983 to 1996, the years in which he was at the peak of his media exposure and popular writing, Sagan was still turning out more than one scientific paper per month. His peers, though, saw him as nothing more credible than 'a publicist' for science. 'What Sagan was most famous for, and what got him in the biggest trouble with the academic establishment, was his Brobdingnagian outpouring of popular articles and interviews,' Shermer says.

After his coming of age in the post-war era, Sagan came to see that science is a tool used for political purposes – and that scientists had largely ignored their responsibilities to make sure that it is used the right way. He became passionately committed to returning science to its proper function of exploring the universe and, where possible, making it a better place to be.

In the early 1970s, environmental scientist James Lovelock wanted to find a way to measure how air moved around the globe. He soon realised that the chlorofluorocarbon (CFC) molecules that were then found in every refrigerator, freezer, can of hairspray or deodorant and myriad other products were a godsend. Once released, CFCs are stable in the atmosphere – they don't break down easily. They are also not naturally occurring: they enter the atmosphere only above populated areas of the Earth. So if you could travel round the globe with a CFC sniffer, you might be able to track the atmospheric currents.

Locking himself away in the laboratory he had built in the garden of his Wiltshire home, Lovelock set about building the most sensitive CFC sniffer the world had ever seen. He called it the electron capture gas chromatograph, and by the time he had finished it, it was sensitive enough to detect concentrations of atmospheric CFC equivalent to a single drop of water in a swimming pool. Ironically, it was so sensitive that he had to make his family stop using products that contained CFCs because they were interfering with his preliminary test results.

When the instrument was ready for service in the wider world, Lovelock booked a passage on the research vessel RRS *Shackleton*, which was making a return voyage to the Antarctic. All the way there and back, he took measurements of the concentrations of CFC in the atmosphere.

After his return, Lovelock attended a conference where he chatted over coffee with a scientist from DuPont, the principal manufacturer of CFCs. Together, the two men idly observed that Lovelock's measurements of the total amount of CFC molecules in the atmosphere tallied almost exactly with the total worldwide production to date. They thought it an interesting coincidence, but nothing more than that. Eventually, though, a chemist called Sherwood Rowland came across this little nugget of information. And he thought it was astonishing.

The news that all the CFCs that had ever been manufactured were still in the atmosphere gave Rowland an idea for a research project. He knew that the CFC molecules would be stable in the lower atmosphere, but he also knew that they would eventually rise into higher layers of the atmosphere and become exposed to increasing levels of solar radiation. This, he reasoned, would break the molecules down into their constituent parts. But what would happen after that? Mario Molina, who was working as a postdoctoral researcher in Rowland's lab, decided that he would be the person to find out. And by Christmas 1973, the full horror of the situation had become clear.

Molina found that the CFCs would take a few decades to reach the stratosphere, the atmospheric layer that sits between 10 and 30 miles above the Earth's surface. Once they arrived there, solar radiation would break them apart, releasing free chlorine atoms. And these chlorine atoms would wreak havoc on the ozone layer.

Ozone is a molecule composed of three oxygen atoms (an ordinary molecule of oxygen contains two). There isn't much of it in the atmosphere. What there is is all in the stratosphere, in extremely low concentrations. If you compressed it all together, the ozone would blanket the Earth's surface in a layer no thicker than tissue paper.

Nevertheless, it does an important job. Ozone absorbs ultraviolet light, shielding the Earth's surface from the harshest of the Sun's rays. Thanks to the ozone layer, we are protected from radiation that induces skin cancer and blindness. According to the World Health Organization, if CFC production had carried on unchecked, the ensuing depletion of the Earth's ozone layer would have caused an extra 500 million cases of skin cancer per year by 2050. Ten years on from that, the figures would have tripled. It is clear that destroying the Earth's ozone would have serious effects on the human race. And it was clear to Molina in 1973 that CFCs *would* eventually cause such destruction.

The ozone molecule is unstable; oxygen is much more stable as a molecule when composed of the usual two atoms. Free chlorine atoms, released from CFCs by solar radiation, would easily knock that extra oxygen atom from an ozone molecule. Together they would form chlorine monoxide, one of the highly dangerous molecules that chemists know as free radicals. Free radicals have a spare, reactive electron in their chemical make-up, and that makes them hungry for something to react with. In the stratosphere, chlorine monoxide would greedily mop up any free oxygen atom, forming a stable oxygen molecule and freeing the chlorine atom to start the reaction all over again. It was a chain reaction, in other words. And Molina knew that was bad news.

His initial reaction, he says, was disbelief – he thought he must have done something wrong in his calculations. But he also says that a chill ran down his spine. He knew that if he was right, then this was dangerous.

Rowland and Molina checked their calculations, discussed them with colleagues and searched for flaws in their analysis. They failed to find any, and in June 1974 they published their findings in *Nature*. A few months later they discussed the results publicly for the first time, at a meeting of the American Chemical Society in Atlantic City. By October, a US Government committee commissioned the National Academy of Sciences to conduct a study into the question of whether the ozone layer really was under threat from human activity.

Not that Rowland was waiting around for the National Academy of Sciences report to come out. The US Environmental Protection Agency calculated that non-malignant skin cancers would rise by 5 per cent for every 1 per cent reduction in stratospheric ozone. Deaths from cancer would increase as fast as ozone concentration decreased. 800,000 tons of CFCs was being released each year, and each chlorine atom they eventually unleashed on the stratosphere was destroying thousands of ozone molecules. It looked as if the ozone layer would eventually be depleted by 20 to 40 per cent. Rowland called for an immediate ban on non-essential CFCs. The industry railed at the possibility – and war was declared.

Unfortunately, most scientists were not willing to fight on Rowland and Molina's side. Some of them even fought for the opposition. In 1975 the Chemical Specialties Manufacturer's Association, a US industry umbrella organisation, brought Richard Scorer, a physics professor at London's Imperial College, to America. His job was to sow seeds of doubt. During his six-week tour, Scorer told anyone who would listen – the viewers of the prime-time TV show *Firing Line*, for example – that tales of ozone destruction were just 'scare stories', and that the supposed impact of CFCs was 'utter nonsense'. The Earth's atmosphere was the 'most robust and dynamic element in the environment. Man's activities have very little impact on it.'

Scorer's tour had no impact on the views of the scientific community, but surveys indicated that it increased by 50 per cent the public's awareness that there was scientific opposition to the claims being made by Rowland and Molina. That was more than enough to keep the controversy alive – and a ban at bay.

By 1976, Rowland was describing himself as 'impatient' for a ban on CFCs. He wasn't alone: at the ironically named 12th International Symposium on Free Radicals, held in Laguna Beach, California, in January of that year, other scientists echoed his concerns that nothing seemed to be happening. But that was not about to change.

When the US National Academy of Sciences issued its report in September 1976, its conclusions were so weak that the next day's *New York Times* reported the Academy as recommending a curb on aerosols, while the headline of the *Washington Post* screamed out 'Aerosol Ban Opposed by Science Unit'. As Lydia Dotto and Harold Schiff, authors of *The Ozone War*, point out, the equivocal nature of the Academy's report meant that both newspapers were 'fundamentally right'.

Things took a long time to improve. Alan Miller, a lawyer working at the Natural Resources Defense Council, has called 1977 to 1985 the 'Dark Years'. Although aerosol sprays using CFCs had been banned in the United States, sales of non-aerosol CFCs

– such as motor vehicle refrigerants – were soaring to new heights. And this was more than a decade after Rowland and Molina had shown CFCs to be profoundly dangerous.

'What's the use of having developed a science well enough to make predictions if, in the end, all we're willing to do is stand around and wait for them to come true?' That was Rowland's outburst to a reporter from *Newsday*. Most of the scientists involved in the ozone wars of the 1970s would describe their attitude as 'properly cautious'. But when the controversy led them to do what one scientist angrily termed 'science in a goldfish bowl', they were crippled by the public exposure, and they reacted badly.

Take the experience of Harvard atmospheric physicist Michael McElroy, for example. Industry representatives singled out his red hair and pale skin as proof that scientists had a 'special interest' in getting CFCs banned. The trade magazine *Aerosol Age* remarked that this man would naturally be an advocate of anything that might cut skin cancers. Even more shocking, perhaps, is this subversive jibe, written by a microbiologist in the pages of *Nature*:

> [O]n the beach at Cape Canaveral in Florida, I saw a red-headed man, sunburned to look like a boiled lobster, applying Novocain cream to his glowing back. The only unusual circumstance was that the man was Mike McElroy, whose field is the physics and chemistry of planetary atmospheres and who has loudly warned us against the ultraviolet perils of destroying the ozone layer … Surely he, of all people, should have kept his shirt on.

In a 1988 interview, McElroy admitted that he made scientific contributions to the debate for a decade without supporting a CFC ban. He was, he said, more concerned about the 'credibility problem' that science was facing because of the 'gaps in our under-standing' than he was about the dangers of depleted ozone. Atmospheric scientist Stephen Schneider took a similar line: he and his colleagues, he says, were 'caught between the exaggerations of the advocates, the exploitations of political interests, the media's penchant to turn everything into a boxing match, and your own colleagues saying we should be above this dirty business and stick to the bench'.

In contrast to this equivocation, Rowland and Molina stuck their necks out and stood up for the ban. Rowland's colleagues shunned him for his activism. Almost no university chemistry departments would have him come and speak for nearly a decade – unthinkable for a chemist of his calibre. Twelve years passed without him being invited to speak to industry groups. Even James Lovelock thought Rowland too rash: he called for a 'bit of British caution' in the face of Rowland and Molina's 'missionary' zeal for a ban on CFCs. Rowland says that taking a political stand over the science of ozone affected his reputation in the scientific community 'on a permanent basis'. Now, he says, be belongs to a group that is 'forever suspect'.

In the end, it was only the terrifying discovery of a hole in the ozone layer over Antarctica that galvanised the scientists. This was the point, McElroy later said, when he decided that it really was 'time to be very serious about regulation'. The hole began to appear in September 1976, just as the National Academy of Sciences issued its equivocating report. But, although everyone was meant to be watching the ozone, no one noticed.

NASA's satellite observation system missed the hole completely. The British observing station on the ground at Halley Bay in Antarctica didn't, but the data it gathered weren't being entered into any computers; instead, they were piling up in a Cambridge laboratory. For four springs (this is the southern hemisphere, remember), the seasonal disappearance of ozone passed all the scientists by. Then, in 1981, some Cambridge students finally got round to inputting the last few years' worth of data.

They didn't take long to notice the anomaly. The most trusted data from the American Amundsen-Scott ground station were indicating 2 to 3 per cent drops in ozone concentration, but according to the British instruments the springtime depletion was reaching 60 per cent. Joe Farman, who was heading the British team, contacted NASA to see whether their satellite had seen the same thing. He received no reply. One of his students got excited, and said they should publish. Farman said no – and tell no one. If they scaremongered, and they turned out to be wrong, all their funding would disappear. Farman decided to wait until he could check that his instruments weren't going awry.

The spectrophotometer at Halley Bay measured the ozone concentration in the sky above by checking which frequencies of light made it through the atmosphere and which didn't. The more UV that made it through, the less ozone there must be in the stratosphere. But Farman's instrument was nearing the end of its life; a replacement was waiting in Cambridge. When that replacement was finally installed and began taking measurements, it found the same seasonal dip in ozone. The 1984 readings showed a 40 per cent dip over a period of around 30 days between September and October. The hole stretched from Halley Bay to a second measuring station a thousand miles to the north-west. It was a big hole, in every sense.

So why hadn't the NASA satellite seen it? A pervasive rumour quickly circulated among the ozone-hunting community: the program NASA was using to analyse the satellite data had thrown out the anomalously low values. Though an appealing idea – *schadenfreude* is alive and well in science, especially when researchers are looking across national and cultural borders – it is not quite true. The NASA satellite receiver marked the data as anomalous: very different from what was expected, and probably the result of error. The anomalies were 'flagged' for checking later. Unfortunately for the NASA researchers, when they got round to it, they checked the anomalies against the readings of the Amundsen-Scott ground station, and its instruments, unlike Farman's, *were* awry. Amundsen-Scott was recording ozone levels nearly twice as large as those the satellite recorded. And since that was more in line with expectation, the NASA researchers relaxed.

This is not to say that they threw the anomalous data out – they just took their time. It was a delay that the team leader, Richard McPeters, is probably still cursing to this day. We saw in Chapter 2 that data can be slippery. We saw in Chapter 6 that being first to a solid result is everything to the scientist. Perhaps that's why McPeters has since claimed to be the first to report the ozone hole, in an abstract submitted in 'late 1984' to the organisers of a Prague conference. But Farman remains the man officially recognised as finding the Antarctic ozone hole. His team sent their findings to *Nature* in December – they arrived in *Nature*'s offices on Christmas Eve – and they were published, to universal astonishment, on 16 May 1985.

According to the historian of science Maureen Christie, the hole could conceivably have been found as early as 1981; the British team 'could have saved two years if the data backlog had not developed, and up to another two if the team leader had been a bit less cautious'. NASA, meanwhile, had set up their system on the assumption that data on Antarctic ozone would be mundane rather than interesting, with no allowance 'for the possibility of surprises'. So it was that, eight years after the hole appeared, the scientists finally had something with which to shock the politicians into action.

Experts at the United Nations Environment Programme now

estimate that the 1987 Montreal Protocol, the international treaty that limited emissions of ozone-depleting chemicals, prevented up to 20 million cases of skin cancer and 130 million cases of eye cataracts. In 2010 they reported that ozone concentrations are no longer decreasing. Though they are not yet increasing, pre-1980 ozone levels are expected to be regained before 2050 over most areas of the Earth. Over the poles, where the most severe depletion occurred, full recovery may take an additional fifty years. In the face of the CFC crisis, no one can pretend that all the inventions of science are an unqualified boon. But at least free, radical scientists can help to solve the problems that progress creates.

In 1963, Dennis Gabor published a book called *Inventing the Future*. It makes fascinating reading now, because Gabor, a Hungarian-born scientist and inventor (of the hologram, as it happens, for which he won the 1971 Nobel Prize in Physics), opens with a startling declaration. 'Our civilisation faces three great dangers,' he says. 'The first is destruction by nuclear war, the second is being crippled by overpopulation, and the third is the Age of Leisure.'

That nuclear war and overpopulation are dangers is not a startling revelation. Gabor's assertion is startling because the idea of a dangerous Age of Leisure seems like a joke. Many of us grew up being told that the best of times were just around the corner, that scientists would be kept busy inventing ways for the rest of us to occupy ourselves while robots and computers took care of everything. But it was no joke to Gabor: this age is 'not yet with us, but it is coming towards us with rapid strides', he wrote. We may now have laughed off the idea of the Age of Leisure, but it is worth noting Gabor's serious tone, because some other twentieth-century pronouncements still have a hold over us. The most enduring is the idea that science is more powerful than nature.

In a 1963 CBS documentary, the chemist Robert White-Stephens looks and sounds every inch the authoritative voice of science. He wears a lab coat, a neat moustache and thick-rimmed spectacles. His phrases are delivered in grave, Churchillian cadences. 'The modern chemist, the modern biologist, the modern scientist believes man is steadily controlling nature,' he says.

White-Stephens was responding to accusations made by a young biologist called Rachel Carson, who had published a book questioning the wisdom of America's new love affair with pesticides. Until 1945, most wars had ended because of insect-borne illnesses such as typhus: too many soldiers were dying of disease for fighting to continue. The invention of dichlorodiphenyltrichloroethane – DDT – changed that, and in the post-war era the chemists cashed in on the kudos they had garnered for themselves. Chemistry, its proponents suggested, could also change peacetime.

It certainly seemed plausible, and so governments invested in giant industrial complexes that churned out tons of chemicals for use in agriculture and city sanitation. US Public Health Department films show DDT being sprayed on happy children eating sandwiches in a public park, on others splashing in the municipal swimming pool, on mothers holding babies while watching community events. The chemists were going to eradicate the insect pest. And, despite the fact that one of the insecticides used, tetraethylpyrophosphate (TEPP), was nothing but the refined essence of a German nerve gas compound, it apparently never occurred to them that these pesticides could harm other organisms too.

It was only when people noticed birds dying that anyone began to express concern. Carson put across the extent of the threat with poetic clarity. The indiscriminate spraying, she said, could quickly lead to a springtime devoid of birdsong: a 'silent spring'. America, she eloquently argued, must rein back its release of chemicals into the environment.

Beset by illness and personal tragedy, Carson took four years to research and write her book. It was published in 1962 to critical acclaim, public alarm and vociferous scorn from many scientists. Emil Mrak, a food scientist and the chancellor of the University of California at Davis, testified to the US Congress that Carson's scientific conclusions were 'contrary to the present body of scientific knowledge'. Loudest among the scientific critics, though, were White-Stephens and his colleagues in the American pesticide industry, backed by a $250,000 anti-Carson war chest. Carson was advancing 'gross distortions of the actual facts completely unsupported by scientific experimental evidence and general practical experience in the field', White-Stephens alleged. 'The real threat, then, to the survival of man is not chemical but biological, in the shape of hordes of insects that can denude our forests, sweep over our croplands.'

Carson's thorough grasp of the science, coupled with the calm demeanour she displayed during her rare public appearances to defend the book – she was fighting a losing battle against cancer – saw off such offensives. The critics were eventually reduced to making crude personal insults: Carson was derided as 'hysterical', 'emotional', a 'tool of Communist menace', and a spinster who could know nothing about safeguarding future generations.

Despite these efforts to discredit and discourage Carson and her supporters (many industry scientists had covertly helped with her research, and those who were openly quoted in the book lost their jobs on its publication), majority opinion gathered behind the message of *Silent Spring*. Spurred into action by public concern, the US Government passed a flurry of laws on environmental protection. Quite rightly, Carson has been called the 'fountainhead of the modern environmental movement'.

Events in science that have had the potential to change humanity so profoundly are rare. Carson's insight seems on a par with

Edward Jenner's invention and championing of vaccination. Until *Silent Spring* was published, few members of the public ever thought that human beings were connected with, or depended on, the environment around them. This was not a result of ignorance, but of scientific arrogance: they believed the assurances of scientists that humans were now in a technological position to take control of nature and harness it for human good. Stewart Udall, then US Secretary of the Interior, remembers this time as the age of 'the atom changing our lives, of the conquest of nature, of technology being the great thing that was going to change the world'. The natural world, he says, was 'pushed into the background'.

That was the spirit in which DDT was sprayed with such astounding abandon on farms, streets, schools, swimming pools and the countryside. *Silent Spring* destroyed that spirit: suddenly, people realised that humans are part of the environment, not standing in isolation above it. Yet despite Carson's extraordinary work, it is a lesson we are still struggling to learn.

The beauty of Rachel Carson's prose is breathtaking – *Silent Spring* was, in part, such a huge success because Carson captured the poetry of nature in her writing. In her last letter to her friend Dorothy Freeman, Carson waxed eloquent about the monarch butterflies of Maine. Carson was in the late stages of her cancer and aware that she was unlikely ever to see the monarchs return to her beloved Maine after their winter migration. She and Freeman had spent that September morning together on the lawn of the Newagen Inn and enjoyed 'the sounds of the wind in the spruces and surf on the rocks, the gulls busy with their foraging, alighting with deliberate grace, the distant views of Griffiths Head and Todd Point, today so clearly etched, though once half seen in swirling fog'. Most of all, though, Carson told Freeman, she would remember watching the monarchs begin their migration: 'that unhurried westward drift of one small winged form after another, each drawn by some invisible force'. Evidently, Carson and Freeman had discussed each butterfly's fate never to return from this closing journey of their lives. The monarchs, Carson said, taught her about the cycles of life: 'it is a natural and not unhappy thing that a life comes to an end'.

Within a few months Carson had died, leaving a legacy – a new sense of environmental responsibility – that we have yet to fully work out. The monarchs are still caught up in that legacy: thanks to shifting weather patterns and the overuse of weedkillers, the inhabitants of Maine and the rest of continental America are seeing fewer monarchs return each year.

It is not necessarily an irreversible trend, as NASA scientist James Hansen will tell anyone who cares to listen. In the summer of 2008, concerned at the monarch's decline, Hansen took his grandchildren out into the wilds of eastern Pennsylvania to find some milkweed, the only plant monarch caterpillars will eat. They dug some up and planted it in his garden. The following year, they found the transplanted milkweed plants dotted with monarch caterpillars. Hansen and his grandchildren now take the seedpods from the milkweed, and plant them around their land in a tiny, near-futile effort to repopulate America with the butterflies that Carson loved so deeply.

But it is only doing nothing at all that is truly futile, Hansen believes. That is why, in 2004, at the age of sixty-three, he embraced life as a climate change activist. Two years later, *Time* magazine named him as one of America's 100 most influential people. That was also the year that Hansen got himself arrested for the first time. Just doing science, he says, is no longer enough.

It's not as if he hasn't done *enough* science. Hansen is one of planetary science's most respected researchers. He has a hugely prominent role as the director of NASA's Goddard Institute for Space Studies, and he is also a professor at Columbia University. He has won numerous awards for his research. Hansen knows, more than anyone, what happens when a planet is in thrall to global warming.

Take a look up into the sky the next time Venus is visible. You won't be looking at its surface: the planet is cloaked in clouds of sulphuric acid and, beneath that, carbon dioxide. Under this thick, stifling atmosphere, the surface of Venus is a barren waste that bakes at temperatures above 450 degrees Celsius. Venus is often referred to as the Earth's twin: its diameter is just 5 per cent less than the Earth's, and it has around four-fifths of the Earth's mass. James Hansen's mission in life is to make sure that those remain the only similarities.

Hansen is one of the scientists who worked out that the searing surface temperature is due not only to Venus's proximity to the Sun, but also to the blanketing effect of the carbon dioxide atmosphere. Knowing this made him deeply concerned about reports that the amount of carbon dioxide in the Earth's atmosphere was growing.

In 1988, the US Congress asked Hansen for an opinion on the 'greenhouse effect'. Some of the Sun's energy that hits our planet is reflected back from the surface. Carbon dioxide and other greenhouse gases absorb some of this reflected energy, preventing it from radiating back into space. A few scientists had begun to warn that increasing amounts of carbon dioxide in the atmosphere would result in rising temperatures on the Earth. If the balance of energy in versus energy out became too skewed in the wrong direction, the atmosphere could eventually heat to disastrously high temperatures.

Hansen's response to Congress was uncompromising. Aware of the claims, he had already stopped studying the atmosphere of Venus and started studying the atmosphere closest to home. 'The greenhouse effect is real,' he told Congress, 'it is coming soon, and it will have major effects on all peoples.' The scientific evidence for this, he said, was 'overwhelming'.

This is not the place to go into the details of claims and counter-claims about global warming (I would recommend Hansen's books for that), but Hansen's testimony to Congress, which stated that human activity is responsible for increased levels of carbon dioxide in the atmosphere, triggered a backlash from the industries, such as power generators and motor vehicle manufacturers, that were linked with carbon dioxide emissions. That backlash, and the political fallout, is continuing.

Right at the centre of the controversy is the IPCC – the Intergovernmental Panel on Climate Change. It is a Nobel Prize-winner: the 2007 Nobel Peace Prize was awarded jointly to the IPCC and Al Gore 'for their efforts to build up and disseminate greater knowledge about man-made climate change, and to lay the foundations for the measures that are needed to counteract such change'. But it is clear to many that the IPCC could do better. The influential physicist and climate activist Joseph Romm summed up the problem thus: 'Most scientists – and the IPCC in particular – have tended to overemphasize uncertainty on the key issues.'

Like the panel on alien communications at the Royal Society's Kavli Centre, the IPCC is keen not to be seen as too troublesome when faced with public scrutiny. For all the bold behaviour of some individuals, gather scientists together so that they are forced to speak with one voice, and they naturally, instinctively, make a concerted effort not to be alarmist, not to say things that might be interpreted as problematic. As a result, the IPCC, a collection of scientists speaking to the governments that control their funding, has underplayed the various impacts of the greenhouse effect – on sea level as glaciers melt, for instance. How? By overemphasising the uncertainty surrounding the data on climate change.

Hansen has pointed out that, although general funding for tackling climate change has increased dramatically in recent years, the lion's share has gone to those who are the most cautious. 'It seems to me that scientists downplaying the dangers of climate change fare better when it comes to getting funding,' he says. He has personal experience of this. In 1981, the US Department of Energy reversed a decision to award his research group a grant. They explicitly told him that it was because they didn't like a paper he had published on the likely effects of continued fossil fuel use.

The trouble is, if a body such as the IPCC plays down the likely impact of global warming, how can anyone decide on the most appropriate response to the real situation? As Carl Sagan pointed out when considering the likely effects of a nuclear winter, if scientists don't make the objective view available, how will anyone know what it looks like and understand what to do? Hansen's response to the conservatism of the IPCC is straightforward: 'Do we not know enough to say more?' he asks. In 2004, he broke a fifteen-year 'self-imposed effort to stay out of the media' and began to speak up.

In 2006, the *New York Times* reported on NASA's attempt to silence Hansen. His call for immediate cuts to carbon emissions, made at a meeting of the American Geophysical Union, led to NASA insisting that Hansen's supervisors stand in for him in any future media interviews. His response was to shrug his shoulders and carry on; to Hansen, this is a moral issue that has as much social import as civil rights or fascism.

When governments drag their feet over such issues, civil disobedience becomes the only option for citizens, Hansen says. That is why, in March 2009, he joined a protest against coal-burning power stations held at the Capitol Power Plant in Washington, DC. The organisers celebrated the event as 'the biggest act of civil disobedience against global warming in American history', and it was there that Hansen first declared his willingness to be arrested for the cause.

He didn't have long to wait. On 23 June, West Virginia State Police arrested Hansen and dozens of other demonstrators, including the actress Daryl Hannah, for trespassing on the property of a coal-mining company. Massey Energy were planning to blow the top off a Raleigh County mountain to get at the coal seams beneath, a practice that has been widely condemned for the environmental havoc it wreaks. In September 2010 Hansen was arrested again, this time during a protest – against the same practice – held outside the White House.

Hansen is always careful to make it very clear that in participating in these protests he is acting in a personal capacity, and not as a NASA representative. But, that done, he shows no sign of toning it down. These days he is advocating putting legal pressure on governments who, he says, have a 'responsibility to protect the rights of young people and future generations'.

Unlike Rachel Carson, James Hansen is not a particularly gifted communicator. His writing is plain, sometimes clunky, and almost entirely untroubled by poetic flourishes. It is quite the kind of writing one would expect of a self-declared 'slow-paced taciturn scientist from the Midwest'. One thing Hansen does have, and what seems to have set him on the same path as Carson, is grandchildren. His biggest fear, as a climate scientist with all the facts before him, is that his grandchildren will one day look back and justly accuse him of understanding exactly what was happening, but doing nothing about it. Helping them raise a few butterflies is not enough.

This raises an obvious question. Plenty of scientists – plenty of climate scientists – have grandchildren. So why does Hansen cut such a lone figure on the scientific barricades? We might put another related question here, too: why did the fight to ban CFCs take so long? Naomi Oreskes and Erik M. Conway framed the same issue another way in their revealing book *Merchants of Doubt*. They delve into the details of some of the biggest scientific battles of the last hundred years and find the scientists strangely disappointing. Oreskes and Conway wanted to tell 'heroic stories of how scientists set the record straight' on acid rain, climate change, tobacco marketing and the ozone crisis. But only in a very few cases were they able to. 'Clearly, scientists knew that many contrarian claims were false,' they point out. 'Why didn't they do more to refute them?'

It is clear that the open anarchy we have seen in the actions of Sagan, Carson, Hansen and others, though inspiring, is rather rare. Its rarity is in marked contrast to the prevalence of, say, scientific fraud. Analyse the history of scientists speaking truth to power, and you will find the scientists strangely timid. Sometimes, it turns out, scientists are nowhere near as anarchic as you might – given all we have seen – reasonably expect.

One reason for this is simple human timidity. Some scientists have been reluctant to make strong claims about climate change lest contrarians attack them. An oceanographer once told Oreskes that she would rather err on the side of caution in her estimates because it made her feel more 'secure'. The threat of personal and professional attacks – intimidation and bullying – has put many scientists off correcting the erroneous outpourings of climate change deniers.

Other reasons have much more to do with the downsides of the secret anarchy of science. There is, for instance, the self-interested desire to 'just get on with it'. In the same way that Barbara McClintock revelled in rejection because she could continue her research untroubled by interested colleagues, many researchers avoid controversy because they want to pursue scientific inquiry and nothing else.

Then there is the fact that scientists cling to the notion that the truth will out in the end. It is not the job of the scientist, some say, to get involved with the day-to-day process of informing the public about the scientific case in matters of public policy. Oreskes and Conway focus on this excuse and conclude, magnanimously, that scientists' failures to engage with pivotal issues mostly arise from a hopeless naivety. Scientists, as a whole, have a rose-tinted view of the power of science and genuinely believe that if they just quietly continue their laboratory research, then the scientific quest for truth will eventually triumph. The secret anarchists have, in other words, fallen victim to their own deception.

The final excuse for inactivity may be the most toxic by-product of the secret anarchy. Many scientists have announced that their expertise is of no value when it comes to deciding on a course of action. In a hearing about ozone depletion held before the US Senate, Michael McElroy said that when it came to making policy recommendations, his own advice 'isn't worth any more than the advice of any informed layman'. In 2008 the climate scientist Susan Solomon took the same stance, telling the *New York Times* that, 'If we as scientists go beyond what we know into our personal opinions and values, we begin to engage in the same sort of personal speculation masquerading as authoritative that we dislike when it is done by the sceptics.'

Though it could be lauded as humility, such reticence has much more to do with the secret anarchy. After decades of executing the post-war policy of keeping a bowed and subservient head – for the sake of Brand Science – scientists just aren't comfortable with raising their voices. Even when the world needs someone to say something.

This attitude is the one that most needs to change, according to Michael Nelson of Michigan State University. Hansen's position, Nelson believes, is the only morally acceptable one that scientists can take. Scientists, he says, have a special responsibility to engage in activism. 'When scientists reject advocacy as a principle, they reject a fundamental aspect of their citizenship,' Nelson has said. 'Rejecting one's responsibility as a citizen is unethical.' Assertions that scientists are only there to lay out the facts are dangerous for all of us:

> I shudder when I think about the implications of stripping scientists – those who might know more about some given topic than anyone else – of their citizenship. I do not think people know what they are saying or implying when they say scientists should not be advocates, or when scientists justify their lack of advocacy or criticize their peers on this basis.

Scientists, as highly informed citizens, have their own peculiar set of responsibilities, especially because their colleagues and professional forebears have, with the post-war rebranding of science, helped to create the problems that good science alone can solve. Carl Sagan put it thus: 'It is the particular task of scientists, I believe, to alert the public to possible dangers, especially those emanating from science or foreseeable through the use of science.' The quote comes from a book, *The Demon-Haunted World*, that Sagan dedicated to his grandson Tonio with these words: 'I wish you a world free of demons and full of light.'

EPILOGUE

Nearly seven years have passed since that enlightening episode with Stephen Hawking's soup. Now I'm watching another scientist eat – and again, it's rather distracting. This time I am in the cafeteria of the Laboratory of Molecular Biology at Cambridge University, and newly crowned Nobel laureate Venkatraman Ramakrishnan has just sat down at the next table. He is tucking heartily into a banana, and I'm wondering whether I should grab the opportunity to interview him before he wanders back to his lab.

With a little effort, I pull my focus back to Michael Fuller, the man sitting at the table with me. Fuller is the technician who built Crick and Watson's famous model of DNA. He is telling me about the years he spent working with Francis Crick, but he can see that I'm not quite giving him my full attention. Fuller, a warm, generous, walking smile of a man, sees my predicament. 'Do you want to go and talk to Venki?' he asks.

I think about it, then dismiss the idea. 'I might catch him later,' I say. But I know that I probably won't bother. Ramakrishnan won't be able to tell me the secret of how to win a Nobel Prize. The scientists themselves rarely can. Anyway, it's not that secret that I am really here to uncover. Ten days after Crick's death in 2004, a British journalist called Alun Rees published a 'scoop' that he had been sitting on for years. Crick, Rees reported, had been high on LSD at the moment when he and James Watson had discovered the structure of DNA.

It is certainly not impossible: Crick was not shy of drugs. In 1967 he signed a letter to the London *Times* – other signatories included Paul McCartney and Graham Greene – that called for a reform of the drug laws. Around the same time he helped found a pressure group seeking to get cannabis legalised. The group was called Soma, after the socially acceptable mind-altering drug that featured in Aldous Huxley's *Brave New World*. Huxley was a great advocate of the use of LSD and mescaline, among other pharmaceuticals.

Rees said in his article that it was while moving in this circle that Crick met Richard Kemp, a young biochemist who went on to develop a new and extremely efficient way to manufacture LSD. According to Rees, Crick had been Kemp's inspiration: Crick had told Kemp that LSD had enabled him to see the structure of DNA, and that all the academics in Cambridge were using it to free their minds. Unfortunately, Rees got this story from a 'friend' of Kemp's. When he asked Crick directly about his LSD use, Rees reported, Crick 'listened with rapt, amused attention … He gave no intimation of surprise. When I had finished, he said: "Print a word of it and I'll sue."'

The truth is hard to uncover now. Rees's evidence that Crick used LSD to discover the secret of life comes third hand, from unreliable sources, and entirely uncorroborated. Neither Crick nor Watson ever made any reference to it. In his biography of Crick, Matt Ridley gives the idea that Crick used LSD to open his mind to the structure of DNA a summary dismissal. According to Ridley, both Crick's widow and the man who supplied the couple with LSD assured him that their first encounter with the hallucinogen came in 1967. What's more, Ridley says, the drug was 'barely available' in the UK in 1953. The notion that the 'then impoverished and conventional Crick would have had access to LSD when it was newly invented in the early 1950s' is implausible, Ridley argues: 'there is simply no evidence for it at all'.

Something about this doesn't quite ring true, though. LSD wasn't 'newly invented' in the 1950s; it was first synthesised in 1938. By 1947 the pharmaceutical firm Sandoz was marketing LSD under the trade name Delysid as a useful drug for psychotherapy. According to David Nichols of Purdue University, who has researched the history of LSD for the Royal Society of Chemistry, Sandoz made it 'readily available to scientific and clinical investigators for medical research' until the early 1960s.

The idea that Crick, a signatory to a public call for the legalisation of cannabis, was 'conventional' is somewhat laughable – he was anything but. When the Queen came to Cambridge to open the new Laboratory of Molecular Biology building in 1962, Crick stayed away in protest: he was staunchly against the monarchy, and some years later he refused a knighthood. Crick's house parties were legendary for their wild drunkenness. And he was an inveterate womaniser. One secretary tells of being chased round the laboratory benches by a randy Crick; when he caught her she had to stab the stiletto heel of her shoe into his foot in order to escape. Then there's the fact that the 'conventional' Crick regularly smoked pot and used LSD later in life. Ridley reports that Crick found LSD's effects 'fascinating'.

All of this is enough to make me slightly suspicious of Cristof Koch's testimony. Koch, a University of California neuroscientist who describes Crick as his 'mentor', says that although their conversations were very wide-ranging, Crick never mentioned taking LSD. 'He told me about a lot of private things, including his parties. But not once about any serious drug use,' Koch says.

I did say they were *secret* anarchists. Which is why I am hoping that Fuller might be able to shed some light on the situation. He worked with Crick and Watson for years, attended some of those parties, and went out to buy the champagne on the day of the Nobel Prize announcement. He saw them in almost every situation they faced – does he know about any LSD use? He shakes his head. 'But,' he says, 'Knowing Francis … I imagine he would have experimented if he'd had the chance.'

I imagine so too. We have seen that scientists will do anything in their pursuit of discovery, and Crick and Watson were certainly getting desperate. The American Linus Pauling was closing in on the structure of DNA. Rosalind Franklin and Maurice Wilkins were bickering, and being slow to come forward with the data Crick and Watson wanted – so slow, in fact, that Crick and Watson stole what they needed from Wilkins' lab. When Wilkins complained, Crick told him to 'cheer up and take it from us that even if we kicked you in the pants it was between friends. We hope our burglary will at least produce a united front in your group.'

The evidence of anarchy piles up. Crick and Watson weren't even meant to be working on DNA: their boss at Cambridge had told them to stop. They gave a shrug, went underground and carried on in secret. And Crick's attitude to scientific propriety is clear from his pronouncements years later. In 1979, amid many accusations that Franklin wasn't accorded enough credit, Crick declared that she didn't have what it takes to be a top-class scientist. She was 'too determined to be scientifically sound and to avoid shortcuts', he wrote in 1979. Soon after that, he repeated his belief in the merits of scientific anarchy. 'First-class scientists take risks,' he said. 'Rosalind, it seems to me, was too cautious.' One more piece of anarchy, a little hallucinogenic help in visualising the structure of DNA, would hardly have made a difference.

It is not as if this would have made Crick a one-off. As we have seen, LSD seeded Kary Mullis's Nobel Prize. Another Nobel laureate, the physicist Richard Feynman, enjoyed marijuana and LSD (but had already done his best work before he tried them). The cosmologist Carl Sagan was also a regular user of cannabis, and describes many experiences of seeing things in a new way when stoned. His insights were so profound to him that he made tape recordings to try to persuade his 'down' self to take them seriously the next day:

> If I find in the morning a message from myself the night before informing me that there is a world around us which we barely sense, or that we can become one with the universe, or even that certain politicians are desperately frightened men, I may tend to disbelieve; but when I'm high I know about this disbelief. And so I have a tape in which I exhort myself to take such remarks seriously. I say 'Listen closely, you sonofabitch of the morning! This stuff is real!'

Sagan was open to the idea that drug-induced experiences would help his research. His best friend was Lester Grinspoon, a professor of psychiatry at Harvard. They used to get high together, and Grinspoon remembers Sagan asking him for his last 'bud' to help him with the following day's research. 'Lester, I know you've only got one left, but could I have it?' Sagan said. 'I've got serious work to do tomorrow and I could really use it.'

It has to be admitted, though, that while Sagan said that pot improved his appreciation of many things – including, oddly enough, potatoes – there is little evidence that his cannabis use had any impact on his scientific work. He describes one occasion when he was able to recall seemingly irreconcilable experimental facts when stoned, and coming up with something that might pull them together, but admitted that it was 'a very bizarre possibility'. He wrote a paper that mentioned the idea. 'I think it's very unlikely to be true,' he later wrote, 'but it has consequences which are experimentally testable, which is the hallmark of an acceptable theory.'

Hardly a moment of blinding revelation, then. But neither do we have any blinding revelation when it comes to Francis Crick's use of LSD in the 1950s. There is no solid evidence, just conflicting testimony and the presence of a personality that would almost certainly have used anything he could to steal a march on his competitors. We are back to that mantra again: anything goes. When I started this project, Feyerabend's idea that 'anything goes' in science seemed like a glimpse of its dark side. Now, having explored the lengths to which scientists will go in the pursuit of discovery, it has become apparent that 'anything goes' is a virtue – the secret of science's success.

Some of the stories about science in this book might be shocking, but hopefully it is now clear that science often progresses in ways that defy our usual notions of what scientists get up to. And the anarchists have made important discoveries. Einstein might never have proved without doubt that $E = mc^2$, but that doesn't mean it isn't true. What's more, our understanding of the interplay between mass and energy helped to bring about the end of the Second World War. Even more important were fortuitous discoveries that led to the Allies' ability to engineer the atomic bomb ahead of the Nazis. The scientists involved can't exactly take credit for those discoveries – they don't know quite how some of them happened – but they grabbed them with both hands and used them to make the world a better place.

Anything goes; science does what it needs to do. Barry Marshall infected himself with a hazardous dose of bacteria because he was frustrated at the suffering of others – his identification of the cause of stomach ulcers was ultimately selfless. Werner Forssmann tricked and lied his way into a hospital operating theatre because he suspected that, if he had access to its equipment, he could improve our understanding of the heart and find ways to treat otherwise untreatable conditions. Stanley Prusiner couldn't prove that prions existed, but he was convinced that the concept would help researchers to fight the ravages of a swathe of brain diseases. If his colleagues objected to the way he went about it, then that was a small price to pay for new insights that might lead to the prospect of cures for Alzheimer's or Parkinson's disease.

Even the fights, the injuries and the injustices have their purpose. If you want to rise to the top, you and your scientific insight have to be bomb-proof. Any big new idea and its proponent both have to survive so much violence, and unseat such strongly rooted predecessors, that, if they make it through to widespread acceptance, we can be as sure as is possible that they are correct. Most of us are the unwitting beneficiaries of this gladiatorial process. That is why we unhesitatingly board aeroplanes or take aspirin: science is trustworthy. But few of us are aware of the cost at which that trust is achieved. The strange thing is that the scientists would rather you remained in the dark.

When James Watson published an autobiography in 1968, Francis Crick and Maurice Wilkins, his co-recipients of the 1962 Nobel Prize, were furious. According to Matt Ridley, that was because *The Double Helix* took readers into 'the messy, competitive, error-strewn, naughty, human business of grappling with ignorance, rather than to describe science as a stately march towards discovery by paragons'.

This same issue was explored by Peter Medawar throughout his writings. 'It is a layman's illusion,' he said, 'that scientists caper from pinnacle to pinnacle of achievement and that we exercise a Method that preserves us from error.' But, for all his brand-busting honesty, Medawar was in no doubt about science's ability to reach those pinnacles:

In terms of the fulfilment of declared intentions, science is incomparably the most successful enterprise human beings have ever engaged upon. Visit and land on the moon? A fait accompli. Abolish smallpox? A pleasure. Extend our human lifespan by at least a quarter? Yes, assuredly, but that will take a little bit longer.

Medawar was being unduly conservative with that last statement. Over the past two hundred years, human life expectancy has doubled in the developed world, thanks to advances in our understanding of healthcare and nutrition. Can we prolong it still further? Quite possibly. Cambridge University's Richard Smith, an expert on population dynamics, points out that each time a natural limit has been suggested, it has been exceeded. In the 1920s, life expectancy in the United States was around fifty-seven years, and the best estimates were that this could be extended by around seven years at most. In 1990, the experts said that, without major breakthroughs in slowing down the rate of ageing, average life expectancy could not exceed eighty-five years. Only six years later, Japanese women left that upper limit behind. Smith wryly notes that the United Nations has now abandoned the practice of estimating upper limits on life expectancy.

Not that these successes of science are problem-free. In a world with a population of seven billion and rising, all kinds of issues, such as food production, housing and healthcare, pose unprecedented challenges. Nonetheless, this is the world that science has created – the world we asked science to create – and the secret anarchists have risen to the occasion. Whether these successes can continue, whether science can solve the next set of problems, will depend on whether we are willing to let the anarchy come out into the open. Science has achieved much during its period of covert action. But can we now, in the light of what we have learned about how science really works and how it has been so misguid-edly rebranded, set up a better system?

Take peer review, for example, today's gold standard for scien-tific publishing. This procedure, where ideas and results are exam-ined by suitably qualified scientists before being published, was not always a standard route to publication. The modern publica-tion system evolved from exchanges of letters between scientists. If one scientist had something to say to another – it never used to be about letting *everyone* know – they would write them a letter. Eventually, as science grew and the letters needed to be distrib-uted to more and more people, the practice of publishing letters for everyone to read was born.

At first, peer review was not part of this system. Einstein was certainly unused to peer review – in the latter part of his career, when peer review had started to become fashionable, he railed against having to modify a paper to meet the objections of his col-leagues before he could publish it. 'I see no reason to address the – in any case erroneous – comments of your anonymous expert,' he wrote to the editor of *The Physical Review*, who had sent one of Einstein's papers to an expert on general relativity. Einstein's objection included a declaration that he had not expected it to be shown to anyone: he had sent the paper for *publication*. 'On the basis of this incident I prefer to publish the paper elsewhere,' he told the editor. And he did. The paper, 'Do Gravitational Waves Exist?', was unquestioningly accepted by another journal, and published complete with the mistake that the reviewer (but not Einstein) had spotted.

The famous Crick and Watson paper on the structure of DNA was also not peer-reviewed prior to publication; the editor of *Nature*, John Maddox, declared its correctness to be 'self-evident'. At the time, the only kind of peer review carried out by *Nature* was done by a member of staff who would take submitted papers with him to the Athenaeum club and, over coffee or luncheon, ask other scientifically qualified members whether the ideas contained in the papers had any merit.

It was the growth in the numbers of professional scientists seeking publication that led to formal peer review becoming the norm. Faced with a barrage of submissions, the journal proprietors simply had to impose a filter. Today, journal editors receive papers from scientists, decide which ones look interesting, and send them out to two or three experts in the field. These experts decide – anonymously, to avoid unpleasantness – whether the papers merit publication. It seems like a sensible system, but only if you believe the misinformation about who scientists are, and how they behave. The fact is, peer review isn't working too well – precisely because scientists are far too human.

Imagine that you are submitting a scientific paper for publication. It will be reviewed by the experts in your field: your competitors. They are not going to reject it just because it's not their work; that would be far too obvious. But the temptations are there. If you have completed work that they are only halfway through, they will be tempted to delay your acceptance – perhaps subconsciously. If your work makes theirs redundant, it will be difficult for them to fall on their sword and admit defeat. If they just don't like your approach, they will be tempted to pick holes in it – or create some. I have heard researchers moan, for instance, about a reviewer who couldn't find flaws in their work, but told the journal editor that the work should be published only if accompanied by this disclaimer: 'The most plausible explanation of these results is that they are somehow wrong.'

Even if reviewers are unbiased and objective, for the system to be effective they need to have the time and the inclination to examine papers thoroughly. Reviewers are humans under enormous pressure. They all know that they can't just refuse to review their peers' work – the journal editors know who they are, after all, and could reciprocate by refusing to even look at papers the reviewers might submit themselves. As scientists will occasionally admit off the record, the over-busy reviewer frequently offers little more than a cursory examination. Present-day peer review does not mimic Michael Faraday's rigorous recreating of every experiment. Some papers are rigorously examined – especially if they make great claims – but not all. Far from it.

The traditional form of peer review is an archaic system. Many scientists admit – privately – that it just doesn't work. Occasionally, they come out and say it in public. Martin Rees, for example, a former President of the Royal Society, has conceded that reviewing by learned journals 'is not the only way to ensure quality control in science'. Electronic publication accessible to anyone, he suggests, would allow scientists to weigh claims, attempt replication and point out – perhaps endorse – papers that merit attention.

Rees is a fan of the preprint archive arXiv.org, an online repository operated by Cornell University. It provides a showcase for new papers in physics and related fields, and most scientists are able to tell at a glance whether a paper merits their attention. If someone were to add to it a recommendation system for registered scientists, something like the review system on the Amazon website, a rival to the decades-old standard peer review system would be born. Yes, it is still open to bullying, but the removal of anonymity would soon halt that in its tracks.

One of the problems the administrators of science have with such a system is that they have built their assessments upon peer-reviewed journals, weighing up a scientist's worth on the basis of how many publications they produce, and in what standard of journal. Here again, we find the cart before the horse: instead of publishing to alert their colleagues to interesting new findings, they are publishing to survive the system and make sure they get enough funding to continue in their line of work.

The fact is, scientists have been the architects of their own problem. For decades they have posed as dependable, trustworthy, non-radicals, and now they wonder why they have a management system that treats them like docile workers on a production line, rather than what they know themselves to be: creative and curious minds, pursuing lines of inquiry that could lead anywhere – or nowhere. Of course, having taken on board the idea that scientists shouldn't cause a fuss in any sphere – part of the tacit agreement of the post-war period – all that scientists can do is moan to one another about the deadening hand of their administrators.

Which brings us to another undesirable consequence of the cover-up. Scientists don't mobilise. They don't agitate. They don't kick up a stink. Through decades of conditioning, rather like wolves who have allowed themselves to be domesticated and slowly bred into yappy chihuahuas, they have been tamed. Scientists, to put it bluntly, have lost their will to bite and snap at anything that lies outside their immediate sphere.

As a consequence, they are a politically inert group who have become convinced that they should advise (if asked), but never seek to influence, the political agenda. 'Scientists should be on tap, but not on top' is how Winston Churchill saw it in – again – the era immediately after the Second World War. It is an ideology that scientists have wholeheartedly, and somewhat cravenly, accepted throughout the decades since. Scientists involved with advising governments have laboured under the self-delusion that they do so on behalf of the wider population. The truth is likely to be a little more self-serving than that: the primary aim has been to appear useful without appearing troublesome (or worse, meddlesome) to those who control science's funding.

If scientists didn't have such a crucial role to play in building and safeguarding our future, that might be acceptable, if all too human, behaviour. The problem is that, for reasons everyone has forgotten, our society now gains next to no input from some of the finest minds in our midst. As Michael Nelson and John Vutevich put it in the *Chronicle of Higher Education*, 'It is a perversion of democracy to muffle the voice of the most knowledgeable among us and consequently amplify the voice of those with the greatest ignorance.'

Thanks to the post-war whitewash, a cloud also hangs over our ethics boards. The very committees that were meant to avoid a repeat of atrocities and murder have, in some cases, caused thousands of deaths through bureaucratic delays. Ethics committees were set up at the same time as scientists and governments were seeking to dispel public fears about the scientist's sense of responsibility, so they were always going to be overcautious. The acts that precipitated the Nuremberg Code took place in extraordinary circumstances that no longer apply (and no amount of regulation on ethics would prevent anyone intent on such acts from carrying them out). Just as 'the Sabbath was made for man, and not man for the Sabbath', ethics committees are meant to serve science. Science is not meant to be enslaved to their ever-widening remit.

The medical literature contains many studies of the performance of ethical review panels that highlight the problems they can cause. A study in Scotland, for instance, approached nineteen committees and found that fifteen of them had designed their own application forms, creating an inconsistent, time-consuming system where applications were subject to the whims and particular interests of those on the committee. Some committees required researchers to submit twenty copies of the documentation, and the time taken for final approval varied from 39 to 182 days – on average, it took three months for researchers to get the go-ahead. Perhaps the most worrying finding was that the final decisions tended to depend on the personal moral stance of the committee members.

It is easy to see why scientists seek ways to sidestep ethics committees. Lives have been lost because of the administrative burden imposed: when one committee delayed a trial designed to test new drugs for heart attacks, around 10,000 people died unnecessarily. As a 2004 editorial in the *British Medical Journal* pointed out, 'The burdens imposed by ethics review might be justified if it could be shown that, on balance, it does more good than harm to patients' interests. Delays may, however, have important consequences and sometimes jeopardise the interests of patients.'

The scientists' attempts to paint their field whiter than white, putting on their 'anxious to please' face, as Jacob Bronowski described it, have contributed to this over-cautiousness. But, in the light of what we now know about the way scientists work, ethics committees would do well to focus on their own flaws. There is very little evidence that scientists are prone to conducting unethical experiments by choice – they have to publish, discuss and defend these results before their peers and the wider public. They also have to look to the next round of grant applications: when the focus is your own career, your own rise to the top of the field, performing experiments that put others in danger is akin to shooting yourself in the foot.

Then there is the issue of science education. How do we inspire the next generation of scientists? Since the 1950s, the public face of science has been dull, spiritless and cautious. Scientists have taken a back seat in society and culture, allowing rock stars, sportsmen and -women, and fame-hungry TV celebrities to win the attention of our children. And we wonder why these naturally curious children, who displayed a delight with science in primary school, show disaffection and lack of interest by the time their eleventh birthdays come around. Once they become aware of what is valued, what is deemed exciting, in the wider world, science loses its lustre. If the high-school students of today were permitted to learn – perhaps through scientists taking a more honest approach with the media – what science and scientists are *really* like, the days of a career in science being the dull, dismal road less travelled would be behind us.

There is also the problem of methodology: science teaching methods and curricula have also been a victim of the cover-up. Children have, by and large, been taught the letter but not the spirit of science. As the philosopher Rousseau suggested, we should not teach children the sciences, but give them an appetite for them.

It is open to question, for instance, whether students really need to learn all of the scientific information on the science curriculum. For most, it is an experience that seems to destroy any interest in science. And anyone who has done a school science practical will know how hard it can be to get results that the textbooks say they should. Why is this seen as a failing? Imagine if teachers were then allowed to use this experience to explain the challenges and rewards involved in making breakthroughs and discoveries, rather than having to press on to the point where the student's notebook contains the 'right' answer. Science teachers have been unwittingly co-opted into the effort to conceal the true nature and spirit of science.

Setting the anarchists free will certainly be difficult while such mindsets prevail. One result has been that many of those who survive their education with sufficient interest in science to pursue it further are of a personality type that perpetuate the problem. They are drawn to science as it has been portrayed: staid and comfortable.

This is a problem that Stanford Ovshinsky, the scientist without a college degree, understands better than most. Traditional forms of education, he says, can hamper scientific creativity in students: 'All the time they're being treated in a "giving of information to you" kind of way, and then when they get out of school they say, "Okay, now you're on your own, think, be creative." After all those years of trying to kill it.' Kary Mullis also worries about the scientists that the post-war growth of the science establishment is continuing to produce. Out of the investment 'came a lot of scientists who were in it for the money because it was suddenly available', he says. These scientists, he observes, were not, like him, 'the curious little boys that liked to put frogs up in the air'.

An interesting question to consider is what kind of science such scientists produce. Surely, if we flood the universities with visionless scientists, it is inevitable that much of science will become boring. Take a 2008 paper by the GEM particle physics collaboration as an arbitrary example. It stretches over twenty pages, has thirty-one authors, and relates to the minutiae of whether a particular kind of subatomic particle called a meson forms a 'bound state' within an atomic nucleus – two decades previously, a couple of physicists had suggested that it might. Unfortunately, the data presented proved nothing; as the last line of the paper states, 'Further data are clearly needed.' It is difficult to tell who would care even if further data *weren't* needed. It is an example of the overspecialised result, a natural inclination of science, and one that has to be resisted where at all possible.

The problem was identified as long ago as 1930, by a Spanish philosopher, José Ortega y Gasset. In order to make progress, science demands that its workers become ever more specialised. The result, Ortega said, is that the majority of scientists are 'shut up in the narrow cell of their laboratory, like the bee in the cell of its hive'. This type of scientist 'is familiar only with one particular science ... in which he himself is engaged in research'. The result, according to Ortega, is a succession of mediocre, tedious advances, not Nobel Prize-winning breakthroughs. And that was pre-war; in the post-war environment things got worse.

In 1950, the German physicist Erwin Schrödinger reprised Ortega's lament. He worried that specialisation creates a societal ennui that could eventually kill the scientific endeavour. 'Never lose sight of the role your particular subject has within the great performance of the tragic-comedy of human life,' Schrödinger warned his fellow scientists. 'If you cannot – in the long run – tell everyone what you have been doing, your doing has been worthless.'

Is such overspecialisation avoidable? Yes, but it demands effort and courage – exactly the kinds of qualities not possessed by those who went into science because it offered an unchallenging route to a secure existence. Andre Geim, who won the 2010 Nobel Prize in Physics, has advice for those wanting to do truly groundbreaking research: don't work where others are working – go off the beaten track. 'If you follow the herd, all the grass is gone,' he says. For a truly significant breakthrough, you 'have to do things no one else is doing. Unless you happen to be in the right place at the right time, or you have facilities that no one else has, the only way is to be more adventurous.'

It seems that, here again, the cover-up may have hurt the progress of science. In the first few decades after the Second World War, scientists opened their field to cohorts of researchers who didn't share their anarchic curiosity. The result was a grudging acceptance of mediocre, tedious advances as equally valid contributions to the scientific endeavour, and a consequent public blindness to science as a vital and fascinating facet of human culture.

Finding a way to keep science outward-looking, relevant, energetic and accessible is vital, Schrödinger said. With uncanny prescience, he pointed out that the 'masses' outside science decide issues such as what gets included in school curricula. According to Ortega and Schrödinger, public disengagement from science through tedium is the first step in a journey that includes, for example, allowing Creationism into the science classroom, and ends with the disappearance of science from popular culture.

The scientists' pose as grey, faceless, unthreatening ushers of a brighter future has had a significant impact. The world is a worse place because of it, and unless something changes, it could get much worse still. That is why we need to set the secret anarchists free.

Here at the Laboratory of Molecular Biology, I have resigned myself to the fact that I am not going to discover the truth about Francis Crick and his use of LSD. His secret – if there is one – is safe. But Crick and Watson are not the only Nobel laureates Michael Fuller has worked with. He arrived at the MRC, aged sixteen, in January 1952 to work as a technician. In the fifty-eight years he has spent here, the centre has won twenty-six Nobel Prizes (it had won three, including Alexander Fleming's, before Fuller arrived). In all his time at the MRC, Fuller has been charged with going out and getting the champagne for each Nobel celebration. Sometimes, he says, Cambridge has almost run dry of bubbly.

I have to ask: what is their secret? Fuller pauses for a long time before he responds. 'Single-mindedness,' he says. 'They can't be deterred by anything or anyone.' There is another pause; he seems unsure whether to make his next remark. 'And big egos,' he adds eventually. 'Incredible egos. They just know, somehow, despite what anyone says, that they are right.'

I am reminded of Albert Szent-Györgyi's comment that scientists are egotists, selfish beings who get their kicks solving the puzzles of nature. It occurs to me, as Venki Ramakrishnan throws his banana peel into the bin and heads back to his lab, that he doesn't look like much of an egotist. But then, as I have said many times, these are *secret* anarchists.

Now that their secret is out, they are not diminished. Having discovered the true depth of their resourcefulness, I am filled with a renewed admiration for the anarchists of science. They make discoveries not despite their humanity, but precisely because of it. If we want more scientific progress, we need to release more rebels, more outlaws, more anarchists. The time has come to celebrate the anarchy, not conceal it.

Science is something we should hold in the highest regard. In the words of Bronowski, 'These are the marks of science: that it is open for all to hear, and all are free to speak their minds in it. They are marks of the world at its best, and the human spirit at its most challenging.' It is challenging because, as he also noted, this reckless, relentless pursuit is made at great personal cost. Claude Bernard once said that the joy of discovery is available only to those who have felt the 'torment of the unknown'. This is science: torment, dreams, visions, restlessness, lying, cheating, despair, brawling, bullying, desperation and – in the end, when everything works out – a moment of euphoria that makes it all worthwhile. Typically, Bronowski found a way to put it more simply. 'Science is the acceptance of what works and the rejection of what does not,' he said. 'That needs more courage than we might think.'

ACKNOWLEDGEMENTS

I would like to thank everyone at Profile Books, in particular Andrew Franklin for his enthusiastic support for this project, for excellent editing, and for pushing me until I stuck my head 'way, way above the parapet'. Thanks also to my agent, Caroline Dawnay, for her assurance that Andrew was not going to be the ruin of me, for directing me to invaluable sources, and for her eagle-eyed suggestions regarding the text.

I am also grateful to the numerous scientists who sent me their papers when I could find no other way to access them, in particular Carol Reeves of Butler University and Laura Manuelidis of Yale University. Thanks are also due to the many scientists who answered my queries, among them David Pritchard, Barry Marshall, Robin Warren, Hans Ohanian and Cristof Koch, and also Stanley Prusiner (who answered my main question by declining my request for an interview). I enjoyed a warm welcome when exploring Cambridge University's Laboratory of Molecular Biology, with Michael Fuller deserving special mention for his radiant honesty and enthusiasm. For pointing me to valuable source materials, I would like to thank Ann Brooks and Adrian Hill.

I had enlightening conversations and encounters with many people during the writing of this book. I can only apologise that of the many I can name only a few here: Alun Rees, Roger Highfield, Jeremy Webb, John Horgan, Cathy Lynn Grossman, Mark Stevenson, Kevin Dutton, Elaine Fox and Charles Ross. Two people deserve special mention: George Lamb and Marc Hughes, who showed me just how engaging and enlightening the humanity of science and scientists can be.

As always, I must acknowledge a debt to the staff of *New Scientist* magazine, who together constitute a fearsome hive mind. And speaking of hive minds, I gained numerous sources and insights through the people I follow on Twitter. They are too numerous for a complete rundown, but a few come to mind and are worth following if you want to continue the explorations begun in these pages. It seems somehow wrong to reduce these hi-tech monikers to old-fashioned patterns of ink on paper, but I am particularly grateful for unwitting help from @AliceBell, @AtheneDonald, @cgseife, @KieronFlanagan, @sciencebase, @sciencecampaign, @sciencegoddess, @tomstandage, @WilliamCB, @xmalik and @ZoeCorbyn.

My immediate family have put up with a catastrophically distracted husband and father over the past year or so; it is only right that I thank them for their patience and support. But I can't promise that it won't happen again.

Michael Brooks
March 2011

NOTES AND SOURCES

PROLOGUE

p. 3   'the Merlins of the Cold War': M. Schrage, 'Physicists' Reign Is Likely to End', *Los Angeles Times*, 3 October 1991.

p. 3   What followed, according to historian Steven Shapin: S. Shapin, *Never Pure: Historical Studies of Science as if It Was Produced by People with Bodies, Situated in Time, Space, Culture, and Society, and Struggling for Credibility and Authority* (Johns Hopkins University Press, 2010). Shapin points to Robert Merton, who has analysed this further. 'A passion for knowledge, idle curiosity, an altruistic concern with the benefit to humanity, and a host of other special motives have been attributed to the scientist,' he says in *The Sociology of Science: Theoretical and Empirical Investigations* (University of Chicago Press, 1973). Merton argues that these are not the natural values of the individual scientists, but of the institutions in which they worked – which demanded the same stance from their employees.

p. 3   'The Stone Age may return on the gleaming wings of Science,' warned Winston Churchill: Speech at Fulton, Missouri, 5 March 1946, published in *Maxims and Reflections* (Eyre & Spottiswoode, 1947), p. 164.

p. 3   Another of Churchill's pronouncements: Address to the Royal College of Surgeons, 10 July 1951, collected in *Stemming the Tide:*

Speeches 1951 and 1952 (Cassell, 1953), p. 91. It is interesting that
Churchill's pre-war pronouncements about science come across
as much more positive. 'The scientific utilisation, by liquefaction,
pulverisation and other processes, of our vast and magnificent
deposits of coal, constitutes a national object of prime importance,'
he said in a Parliamentary debate in April 1928. Talking about Fritz
Haber's invention of an ammonia synthesis method in 1918 (the
invention was crucial for the production of explosives and fertilisers),
Churchill said, 'It is a remarkable fact, and shows on what obscure
and accidental incidents the fortunes of possibly the whole world
may turn in these days of scientific discovery.'

p. 4   Churchill would also have known of Allied scientists testing nerve
gas and mustard gas on their own soldiers: R. Evans, 'Military
Scientists Tested Mustard Gas on Indians', *Guardian*, 1 September
2007, available at http://www.guardian.co.uk/uk/2007/sep/01/india.
military

p. 4   As the renowned biologist Jacob Bronowski put it just a few years
after Hiroshima: J. Bronowski, *The Common Sense of Science*
(Heinemann, 1951), p. 145. It is also worth noting that Bronowski
repeatedly warned his colleagues of the consequences of the public's
distrust of science. In 1956 he wrote, 'People hate scientists ... the
scientist is forced, by the hatred of public opinion, to side with
established authority and government. He becomes a prisoner of the
hatred of the lay public, and by that becomes the tool of authority.' (J.
Bronowski, 'The Real Responsibilities of the Scientist', *Bulletin of the
Atomic Scientists*, January 1956, p. 10.)

p. 4   'very much the image of science that the high-ups in the Royal
Society wanted to put across': T. Boon speaking in *Mad and Bad: 60
Years of Science on TV*, Pioneer Productions for BBC, first broadcast
15 December 2010. Timothy Boon's book *Films of Fact: A History
of Science in Documentary Films and Television* (Wallflower, 2007)
is a treasure trove of material on British science's relationship with
the BBC in the post-war period. Boon cites various memos and
communications where science establishment figures try to influence
broadcast subjects away from the 'perils and dilemmas' angle that so
obviously interested the journalists of the time: 'Can we sometimes
forget war and atomic weapons, industrial advance or productivity
... and say something more of the history and growth of science, of
the great solution wrought by the introduction of the experimental
method ... ?' Or, 'The evil wrought by science springs, not from any

intrinsic evil in science itself, but from its misuse by men who do not really understand what science is …'

p. 4 'You scientists, you kill half the world, and the other half can't live without you': This is from Episode 6 of *A for Andromeda*, which was written by Fred Hoyle and John Elliott and first broadcast in 1962. Hoyle was, in fact, a great scientist as well as a great science fiction writer, but he was not one to toe the party line. That may account for his being passed over for the 1983 Nobel Prize. As one obituarist put it, 'why Hoyle was not included in this award remains a mystery hidden in the confidential documents of the Royal Swedish Academy' (obituary available at http://www.guardian.co.uk/news/2001/aug/23/guardianobituaries.spaceexploration).

p. 5 By 1957, 96 per cent of Americans said they agreed with the statement: Quoted in 'Public Attitude Toward Science Is Yes, but—', *Science*, vol. 215, p. 270 (1982).

p. 5 According to the US Office of Technology Assessment, the average science professor trains around twenty PhD scientists: 'Federally Funded Research', US Office of Technology Assessment, p. 219, available at www.fas.org/ota/reports/9121.pdf

p. 5 'the postures we choose to be seen in': P. Medawar, *Induction and Intuition in Scientific Thought* (Methuen, 1969), p. 26.

p. 5 Feyerabend published a book called *Against Method*: P. Feyerabend, *Against Method* (New Left Books, 1975). The *Stanford Encyclopedia of Philosophy* contains some entertaining anecdotes about Feyerabend: http://plato.stanford.edu/archives/win2009/entries/feyerabend

p. 6 Feyerabend was soon declared the 'worst enemy of science': T. Theocharis and M. Psimopoulos, 'Where Science Has Gone Wrong', *Nature*, vol. 329, p. 595 (1987). This essay by two physicists was prompted by financial strictures placed on British science; the blame was laid at the feet of four philosophers of science.

p. 7 'nothing short of deliberate fraud': R. Westfall, 'Newton and the Fudge Factor', *Science*, vol. 179, p. 751 (1973).

p. 7 taunting his colleagues about his 'secret knowledge': Newton wrote many of his papers in a code that remained undeciphered until John Maynard Keynes worked out the cipher in the 1930s.

p. 9 The ensuing investigation eventually cleared the scientists involved: F. Pearce, 'Climategate Inquiry: No Deceit, Too Little Cooperation', *New Scientist*, 7 July 2010, available at http://www.newscientist.com/article/dn19143-climategate-inquiry-no-deceit-too-little-cooperation.html

p. 9   In February 2010, a poll commissioned by the BBC:
        'Climate scepticism "on the rise", BBC poll shows', BBC
        online news, 7 February 2010, available at http://news.bbc.
        co.uk/1/hi/sci/tech/8500443.stm. A good source on the
        issues associated with climate polls is climatesock.com:
        see, for example, http://www.climatesock.com/2011/01/
        what-do-we-do-when-two-good-polls-say-opposite-things/

p. 9   'People now find it conceivable that scientists cheat and
        manipulate': F. Pearce, '"Climategate" Was "a Game-Changer"
        in Science Reporting, Say Climatologists', Guardian, 4 July 2010,
        available at http://www.guardian.co.uk/environment/2010/jul/04/
        climatechange-hacked-emails-muir-russell

p. 9   A study carried out in March: M. Schwartz, 'Majority of Americans
        Continue to Believe that Global Warming Is Real', available at http://
        woods.stanford.edu/research/majority-believe-global-warming.html

p. 9   This was confirmed in June: Cardiff University, 'Public Perceptions
        of Climate Change and Energy Futures in Britain', available at http://
        www.cf.ac.uk/psych/home2/docs/UnderstandingRiskFinalReport.
        pdf; Yale Project on Climate Change Communication,
        'Climate Change in the American Mind: Americans' Global
        Warming Beliefs and Attitudes in June 2010', available at
        http://www.climatechangecommunication.org/images/files/
        ClimateBeliefsJune2010(1).pdf; Stanford University's 'Global
        Warming Poll' is available at http://woods.stanford.edu/docs/surveys/
        Global-Warming-Survey-Selected-Results-June2010.pdf

p. 9   February's increase in climate scepticism had died away: J. Krosnick,
        'The Climate Majority', New York Times, 8 June 2010. It's worth noting
        that there was a great deal of controversy over Krosnick's claims – see,
        for example, http://pollingmatters.gallup.com/2010/06/reflections-
        on-jon-krosnicks-global.html

p. 10   'the expert and the god': J. Bronowski, The Common Sense of Science,
        p. 142.

p. 10   we will 'know the mind of God': S. Hawking, A Brief History of Time
        (Bantam, 1988), p. 193.

p. 13   'Nearly all scientific research leads nowhere ...': P. Medawar,
        Induction and Intuition in Scientific Thought (Methuen, 1969), p. 31.

p. 14   '... there is nothing underneath our feet': D. Sarewitz, Frontiers of
        Illusion (Temple University Press, 1996), p. 15.

CHAPTER 1

p. 15 'our main "discovery" was the Earth': The quote comes from a letter
that William Anders, the photographer, wrote to *New Scientist*:
'Seeing the Earthrise', *New Scientist*, 24 December 2005, p. 27.

p. 15 still proud of this achievement: Brand tells the story in 'Why Haven't
We Seen the Whole Earth?', *Whole Earth Catalog*, Fall 1969, p. 168.

p. 17 'the most influential environmental photograph ever taken': The
quote is from wilderness photographer Galen Rowell's commentary
in *100 Photographs That Changed the World* (*Life* Magazine, 2003).

p. 18 '... he is wondering how to conceal the fact that he has no opinion
to declare': P. Medawar, *Induction and Intuition in Scientific Thought*
(Methuen, 1969), p. 11.

p. 18 'he made his profound discoveries in the manner of a mystic': H.
Ohanian, *Einstein's Mistakes* (Norton, 2008), p. 3.

p. 18 'far beyond the capacity of human thinking': Quoted in G. Holton,
The Scientific Imagination: Case Studies (Cambridge University Press,
1978), p. 7.

p. 18 '... only to a small degree in this manner': A. Einstein, 'Induction and
Deduction in Physics', *Berliner Tageblatt*, 25 December 1919.

p. 18 'DNA chains coiled and floated': Mullis tells the story in his
autobiography *Dancing Naked in the Mind Field* (Pantheon, 1998), p.
3.

p. 20 'I was down there with the molecules ...': Mullis's quote (and
the one that follows) is from the BBC Horizon documentary
Psychedelic Science. It is available at http://www.youtube.com/
watch?v=j2WurhYEQyY

p. 21 Steve Jobs ... experience with LSD: reported in J. Markoff, *What
the Dormouse Said: How the 60s Counterculture Shaped the Personal
Computer* (Viking Adult, 2005), p. xix.

p. 21 Crick was 'fascinated by its effects': M. Ridley, *Francis Crick:
Discoverer of the Genetic Code* (Harper Perennial, 2008), p. 156.

p. 21 most of the pioneers of Silicon Valley were regular users: W. Kirn,
'Valley of the Nerds', *GQ*, July 1991, p. 96. Text available at http://www.
marijuanalibrary.org/GQ_Valley_of_the_Nerds_91.html. See also A.
Harrison, 'LSD: The Geek's Wonder Drug?' at *Wired* online: http://
www.wired.com/science/discoveries/news/2006/01/70015

p. 21 Every one of them answered yes to both: This and the following quote
are from R. Abraham, 'Mathematics and the Psychedelic Revolution',

MAPS, Spring 2008, p. 6, available at www.maps.org/news-letters/v18n1/v18n1-MAPS_8–10.pdf

p. 22   In April 2008, the evolutionary biologist Jonathan Eisen: http://phylogenomics.blogspot.com/2008/04/what-is-so-bad-about-brain-doping.html

p. 23   'If any ideas have a claim to be called creative …': J. Bronowski, *The Common Sense of Science* (Heinemann, 1951), p. 148.

p. 23   In an article entitled 'Professor's Little Helper': B. Sahakian and S. Morein-Zamir, 'Professor's Little Helper', *Nature*, vol. 450, p. 1157 (2007).

p. 23   A full 20 per cent of them: B. Maher, 'Look Who's Doping', *Nature*, vol. 452, p. 674 (2008).

p. 23   get scientists through writing laborious grant proposals: 'Food for Thought', *Nature*, 31 January 2008, available at http://www.nature.com/nature/journal/v451/n7178/edsumm/e080131–03.html

p. 23   'his pre-eminence is due to his muscles of intuition': the quote is from a lecture delivered, after Maynard Keynes' death, by his brother in 1942. The text of 'Newton, the Man' is available at http://www-history.mcs.st-and.ac.uk/Extras/Keynes_Newton.html

p. 24   '… potential forms of consciousness entirely different': W. James, *The Varieties of Religious Experience: A Study in Human Nature* (1902). The quote is on p. 283 of the 2008 Arc Manor edition, available through Google books.

p. 25   '… but I was unable to decipher the scrawl': This and the quote that follows are from O. Loewi, 'An Autobiographical Sketch', *Perspectives in Biology and Medicine*, vol. 4, p. 17 (1960).

p. 26   According to Henry Dale: H. Dale, 'Otto Loewi', *Biographical Memoirs of Fellows of the Royal Society*, vol. 8, p. 80 (1962).

p. 26   'one of the most remarkable narratives of scientific discovery': A. Lightman, 'Scientific moments of truth', *New Scientist*, 19 November 2005, p. 36.

p. 26   'This was the origin of the "Structural Theory"': quoted in R.M. Roberts, *Serendipity: Accidental Discoveries in Science* (Wiley, 1989), p. 77.

p. 27   He saw himself running alongside a beam of light: A. Einstein, *Autobiographical Notes* (Open Court, 1979), p. 49.

p. 29   'This form of knowledge is pleasing to the erudite …': G. Cardano, *The Book of My Life* (NYRB Classics, 2002), p. 214.

p. 29   The strange story of the genesis of the electric motor: Tesla tells this story in *My Inventions: The Autobiography of Nikola Tesla*, first published in 1919, and widely available in various editions and on Google books.

p. 31 'I didn't need any drawings; the whole plan was perfectly clear in my head': This quote comes from a personal interview with Snyder, conducted by Roger Shephard and extracted in 'Externalization of Mental Images', a chapter from B. Randhawa, *Visual Learning, Thinking, and Communication* (Academic Press, 1976).

p. 31 a carbon copy of the American prototype: This is clear from the report into Russian spying on British atomic research. See Annex J of the Mitrokhin Inquiry Report: http://www.archive.official-documents.co.uk/document/cm47/4764/4764-axj.htm

p. 32 Fermi felt that he didn't want to use the lead block: the story is told in S. Chandrasekhar, *Truth and Beauty in Science* (University of Chicago Press, 1987), p. 21.

p. 33 'feeling of guilt about suppressing the part chance and good fortune played': A. Hodgkin, 'Chance and Design in Electrophysiology: An Informal Account of Certain Experiments on Nerve Carried Out Between 1934 and 1952', *Journal of Physiology*, vol. 263, p. 1 (1976).

p. 33 The British mathematician Paul Dirac: quoted in S. Chandrasekhar, *Truth and Beauty in Science*, p. 20.

p. 33 Michael Faraday apparently felt the same: sources for Faraday's fascinating and instructive story are G. Cantor, *Michael Faraday, Sandemanian and Scientist* (Macmillan, 1991); and S. Chatterjee, 'Michael Faraday: Discovery of Electromagnetic Induction', *Resonance*, March 2002, available at http://www.ias.ac.in/resonance/Mar2002/Mar2002p35–45.html

p. 35 you can hand scientists all the evidence they need: the classic example is B. Barber and R. Fox, 'The Case of the Floppy-Eared Rabbits,' *American Journal of Sociology*, vol. 64, p. 128 (1958).

p. 38 Copernicus ... referred to nature as 'God's Temple': quoted in G. Holton, *The Scientific Imagination*, p. xi.

p. 38 'I began to think whether there might not be a Motion ...': Harvey's idea is laid out in W. Harvey, *On the Motion of the Heart and Blood in Animals*, translated by G. Keynes (The Classics of Medicine Library, L.B. Adams, 1978). Available at http://www.fordham.edu/halsall/mod/1628harvey-blood.html. There are good accounts of Harvey's work available on the web, e.g. http://physiologyonline.physiology.org/content/17/5/175.full

p. 39 'How come that number isn't zero?': de Grasse Tyson asked this question at 'Beyond Belief: Science, Reason, Religion & Survival', a conference that took place at the Salk Institute, La Jolla, California on 5–7 November 2006. There is video available at http://thesciencenetwork.org/programs/beyond-belief-science-religion-reason-and-survival (Session 2).

p. 39   mutually 'alien categories of knowledge': R. Highfield, 'Disbelief proves to be a constant among scientists', *Daily Telegraph*, 3 April 1997, p. 4.

p. 40   Kekulé suggested to his colleagues that … they should 'learn to dream!': Kekulé was speaking at a conference (Benzolfest) in 1890; the transcript is in *Berichte der Deutschen Chemischen Gesellschaft*, vol. 23, p. 1302 (1890).

p. 40   'The eternal mystery of the world is its comprehensibility': A. Einstein, 'Physics and Reality', *Journal of the Franklin Institute*, vol. 221, p. 349 (March 1936).

CHAPTER 2

p. 42   The anarchists won the battle: This story was reported worldwide; see for example the *Time* story at http://www.time.com/time/world/article/0,8599,1703692,00.html. The *Guardian*'s account is at http://www.guardian.co.uk/world/2008/jan/16/catholicism.internationaleducationnews

p. 42   a speech delivered by the Pope back in 1990, when he was Cardinal Ratzinger: Contained in Ratzinger, *A Turning Point for Europe? The Church in the Modern World* (Ignatius Press, 1994), p. 76. A digest of the speech is available at http://ncronline.org/node/11541

p. 42   The sixty-seven signatories made their feelings very clear: 'Newsmakers: On Campus', *Science*, vol. 319, p. 353 (2008).

p. 43   '… much more faithful to reason than Galileo himself': P. Feyerabend, *Against Method* (Verso, 1993), p. 125.

p. 43   According to Giorgio Israel: a translation of Israel's article is at http://freeforumzone.leonardo.it/discussione.aspx?idd=354494&p=156

p. 44   one in five of its accepted papers contained 'questionable data': Editorial: 'Beautification and Fraud', *Nature Cell Biology*, vol. 8, p. 101 (2006).

p. 44   learned 'not to place too much reliance on any single piece of experimental evidence': This quote and the one from Watson that follows are from F. Crick, *What Mad Pursuit: A Personal View of Scientific Discovery* (Weidenfeld & Nicolson, 1989), p. 59.

p. 44   Ptolemy: in the second century AD, he manipulated data: See R.R. Newton, *The Crime of Claudius Ptolemy* (Johns Hopkins University Press, 1977), p. 80.

p. 46   a 'useless fiction': M.A. Finocchiaro (editor and translator), *The Galileo Affair: A Documentary History* (University of California Press, 1989), p. 128. The full quote refers to those who 'resort to useless chimeras such as motions of the Moon and other fictions, without ever thinking about considering the different lengths and depths of the seas.'

p. 47   '... no one can manipulate the fudge factor quite so effectively ...': This quote and the one in the next paragraph are from R. Westfall, 'Newton and the Fudge Factor', *Science*, vol. 179, p. 751 (1973).

p. 47   Ptolemy has been forgiven as 'honestly motivated': the quote comes from an article 'Science: The Ptruth About Ptolemy' in *Time*, 28 November 1977, available at http://www.time.com/time/magazine/article/0,9171,919182,00.html

p. 47   No lesser a figure than Einstein has exonerated Galileo: A. Einstein, Foreword to Galileo Galilei, *Dialogue Concerning the Two Chief World Systems – Ptolemaic & Copernican* (University of California Press, 1967).

p. 48   'many of the risk factors for misconduct also seem to be what makes for good science': J. Giles, 'Breeding Cheats', *Nature*, vol. 445, p. 242 (2007).

p. 48   'scientists who fall deeply in love with their hypothesis ...': P. Medawar, *Advice to a Young Scientist* (Harper & Row, 1979), p. 39.

p. 48   Sources for the Robert Millikan section: G. Holton, *The Scientific Imagination* (Cambridge University Press, 1978); T. Datta et al., 'A New Look at the Sub-Electron Controversy of Milikan [sic] & Ehrenhaft', http://arxiv.org/abs/physics/0507170; Robert A. Millikan, *Oil Drop Experiment Notebooks*, Archives, California Institute of Technology, available at http://caltechln.library.caltech.edu/8/; D. Goodstein, 'In Defense of Robert Andrews Millikan', *Engineering & Science*, no. 4, p. 30 (2000), available at eands.caltech.edu/articles/Millikan%20Feature.pdf; K. Gottfried, M. Peltier and G. Cohen, with a reply by R. Lewontin, 'On Fraud in Science: An Exchange', *New York Review of Books*, 10 February 2005, available at http://www.nybooks.com/articles/archives/2005/feb/10/on-fraud-in-science-an-exchange/

p. 51   'I did not like this, but I could see no other way out ...': H. Fletcher, 'My Work with Millikan on the Oil-Drop Experiment', *Physics Today*, June 1982, p. 43.

p. 54   'I cannot interpret Millikan's italicized statement as anything other than a lie': C. Whitbeck, 'Truth and Trustworthiness in Research', *Science and Engineering Ethics*, vol. 1, p. 403 (1995), available at http://www.onlineethics.org/Topics/RespResearch/ResEssays/cw2.aspx

p. 54   'it cannot be claimed that it has been finally decided': O. Chwolson, quoted in G. Holton, *The Scientific Imagination: Case Studies* (Cambridge University Press, 1978), p. 28.

p. 54   'The beauty here lies with the experiment not the experimenter': G. Johnson, *The Ten Most Beautiful Experiments* (Vintage, 2009), p. 155.

p. 55   intuition is 'an important, and perhaps in the end a researcher's best, guide …': F. Grinnell, 'Misconduct: Acceptable Practices Differ by Field', *Nature*, vol. 436, p. 776 (2005).

p. 55   '… questionable behaviour that threatens the integrity of science': B. Martinson, 'Scientists Behaving Badly', *Nature*, vol. 435, p. 737 (2005).

p. 56   Sources for the Einstein section: H. Ohanian, *Einstein's Mistakes* (Norton, 2008); A. Folsing, *Einstein* (Penguin, 1998); S. Weinberg, 'Einstein's Mistakes', *Physics Today*, November 2005, p. 31.

p. 59   '… fix your attention on their deeds': Quoted in P. Medawar, *Induction and Intuition in Scientific Thought* (Methuen, 1969), p. 10.

p. 59   '… one should not allow oneself to be too discouraged': P. Dirac, 'The Evolution of the Physicist's Picture of Nature', *Scientific American*, May 1963, vol. 206, p. 53. Available at http://www.scientificamerican.com/blog/post.cfm?id=the-evolution-of-the-physicists--pi-2010-06-25

p. 60   'counterfeiting the coin of science': D. Goodstein, 'Conduct and Misconduct in Science', available via his web page at http://www.its.caltech.edu/~dg/conduct_art.html

p. 61   '… A piece of chalk was his only tool': W. Laurence, 'Einstein Offers New View of Mass-Energy Theorem', *New York Times*, 29 December 1934, available at http://select.nytimes.com/gst/abstract.html?res=F30A1EFA3B5D167A93CBAB1789D95F408385F9&scp=1&sq=einstein&st=p. See also D. Topper and D. Vincent, 'Einstein's 1934 Two-Blackboard Derivation of Energy-Mass Equivalence', *American Journal of Physics*, vol. 75, p. 978 (2007), available at muj.optol.cz/~richterek/data/media/ref_str/topper2007.pdf

p. 62   Not until 1949, when he published an autobiography: A. Einstein, *Autobiographical Notes*, Open Court (1979).

p. 62   Sources for the Eddington section: M. Stanley, *Practical Mystic: Religion, Science, and A.S. Eddington* (University of Chicago Press, 2007); H. Collins and T. Pinch, *The Golem: What You Should Know*

About Science (Cambridge University Press, 1993); W. Gratzer, *Eurekas and Euphorias* (Oxford University Press, 2002); H. Ohanian, *Einstein's Mistakes* (Norton 2008).

p. 69   'Observation and theory get on best when they are mixed together ...': Quoted in P. Wood, *Science and Dissent in England, 1688–1945* (Ashgate Publishing, 2004), p. 249.

p. 70   Raymond De Vries and his colleagues called their new paper 'Normal Misbehavior': R. De Vries et al., 'Normal Misbehavior', *Journal of Empirical Research on Human Research Ethics*, vol. 1, p. 43 (2006).

p. 71   The introduction of ethics classes for researchers at the University of Texas: This and the following example come from E. Marshall, 'How Prevalent Is Fraud?', *Science*, vol. 290, p. 1662 (2000).

p. 71   Barbara Redman and Jon F. Merz published a rather remarkable piece of work: B. Redman and J. Merz, 'Do the Punishments Fit the Crime?', *Science*, vol. 321, p. 775 (2008).

p. 72   'Most cases are probably not publicized ...': R. Smith, 'Research Misconduct: The Poisoning of the Well', *Journal of the Royal Society of Medicine*, vol. 99, p. 232 (2006).

p. 73   The science writer Simon Singh: S. Singh, 'Shadows of Days Gone By', *Times Higher Education Supplement*, 6 August 1999, available at http://www.simonsingh.net/1919_Eclipse.html

p. 73   'it is worth remembering that history has vindicated Millikan': D. Goodstein, 'In Defense of Robert Andrews Millikan'.

p. 73   Heredity pioneer Gregor Mendel's data are suspiciously clean: M. Franzen et al., 'Fraud: Causes and Culprits as Perceived by Science and the Media', *EMBO Reports*, vol. 8, p. 3 (2007).

p. 74   '... one of the tiny handful of supreme geniuses ...': R. Westfall, *Never at Rest: A Biography of Isaac Newton* (Cambridge University Press, 1980), p. ix.

p. 74   If young scientists were to find out what really happened in the history of their subject: S. Brush, 'Should the History of Science Be Rated X?', *Science*, vol. 183, p. 1164 (1974).

CHAPTER 3

p. 75   '... stealing our faeces from the latrines to perform sorcery?': S. Lindenbaum, *Kuru Sorcery: Disease and Danger in the New Guinea Highlands* (McGraw-Hill, 1978), p. 101.

p. 76   '... several years before his mother developed kuru herself': You can see the photos in Gajdusek's Nobel Lecture, 'Unconventional Viruses and the Origin and Disappearance of Kuru' (13 December 1976), available at http://nobelprize.org/nobel_prizes/medicine/laureates/1976/gajdusek-lecture.html

p. 77   'Patients know they are to die': D. Gajdusek, 'Kuru: Clinical, Pathological and Epidemiological Study of an Acute Progressive Degenerative Disease of the Central Nervous System Among Natives of the Eastern Highlands of New Guinea', *American Journal of Medicine*, vol. 26, p. 442 (1959).

p. 77   '[T]hey know damn well that we do nothing for the disease ...': The letters are quoted in C. Spark, 'Learning from the Locals: Gajdusek, Kuru and Cross-Cultural Interaction in Papua New Guinea, *Health and History*, vol. 7, p. 80 (2005).

p. 78   Gajdusek's breakthrough paper was published in 1966: D. Gajdusek et al., 'Experimental Transmission of a Kuru-like Syndrome in Chimpanzees', *Nature*, vol. 209, p. 794 (1966).

p. 78   That result was published in 1968: C.J. Gibbs et al., 'Creutzfeldt–Jakob Disease (Subacute Spongiform Encephalopathy): Transmission to the Chimpanzee', *Science*, vo. 161, p. 388 (1968).

p. 78   one of his patients was suffering a slow and agonising death: Prusiner describes his entrance into the field in S. Prusiner, 'The Prion Diseases', *Scientific American*, vol. 272, p. 48 (1995), available at www.mad-cow.org/~tom/prionSP.html

p. 79   he came across what he later called an 'astonishing' report: T. Alper et al., 'Does the Agent of Scrapie Replicate Without Nucleic Acid?', *Nature*, vol. 214, p. 764 (1967).

p. 80   The English mathematician John Stanley Griffith had come up with an answer: J. Griffith, 'Self-Replication and Scrapie', *Nature*, vol. 215, p. 1043 (1967).

p. 81   Equipped with what he terms 'the optimism of youth' and a 'cocky' nature: Prusiner describes his young self as 'cocky' in G. Taubes, 'Nobel Gas: Sure, Stanley Prusiner Deserves a Prize – for His Persistence, Not for His Prions', *Slate*, 11 October 1997, available at http://www.slate.com/id/2096/

p. 81   Prusiner made the claim in a paper published in the journal Science: S. Prusiner, 'Novel Proteinaceous Infectious Particles Cause Scrapie', *Science*, vol. 216, p. 136 (1982).

p. 81 'three decades of investigation have yielded no direct experimental proof': S. Supattapone, 'What Makes a Prion Infectious?', *Science*, vol. 327, p. 1091 (2010).

p. 81 'At every crossroads on the road that leads to the future ...': the quote is borrowed from Belgian playwright Maurice Maeterlinck.

p. 82 represents 'a triumph of the scientific process over prejudice': S. Prusiner, *Prion Biology and Diseases* (Cold Spring Harbor Laboratory Press, 1999), p. 126.

p. 82 they 'wanted to stick close to the evidence': Interview by Carol Reeves, in '"I Knew There Was Something Wrong with That Paper": Scientific Rhetorical Styles and Scientific Misunderstandings', *Technical Communication Quarterly*, vol. 14(3), p. 267 (2005). This is an extraordinary paper, well worth reading, and is studied in more depth later in this chapter. It is available at sullivanfiles.net/science_rhetoric/style/reeves_something_wrong.pdf

p. 82 Richard Kimberlin of the Animal Research Centre in Edinburgh countered it in Nature: R. Kimberlin, 'Scrapie Agent: Prions or Virinos?' *Nature*, vol. 297, p. 107 (1982).

p. 83 'From our point of view, there is no doubt.': L. Altman, 'U.S. Scientist Wins Nobel for Controversial Work', *New York Times*, 7 October 1997.

p. 83 the result could be 'tragic': Chesebro's press release is reported in R. Rhodes, 'Pathological Science', *New Yorker*, 1 December 1997, available at http://www.vanderbilt.edu/AnS/physics/brau/Retirement/Prusiner%20reading/Pathologicalscience.pdf

p. 83 Laura Manuelidis made a similar point: G. Kolata, 'Eye on the Nobel; They Should Give a Prize for Ambition', *New York Times*, 12 October 1997. The 'cold fusion of infectious diseases' quote is from the same source.

p. 84 'Clearly, we are in the very early stages of exploration ...': B. Chesebro, 'BSE and Prions: Uncertainties About the Agent', *Science*, vol. 279, p. 42 (1998).

p. 84 'set off a firestorm' and unleashed a 'torrent of criticism': Prusiner's autobiography is available at http://nobelprize.org/nobel_prizes/medicine/laureates/1997/prusiner-autobio.html

p. 85 'And by then,' Petterson told Reuters, 'it was too late': B. Goldsmith, 'Nobel Winner Could Have Prevented "Mad Cow"', *Reuters World Report*, 6 October 1997, text available at http://www.mad-cow.org/Nobel.html#winner

p. 85 'They are always vulgar, and often convincing.': L. Manuelidis, 'A 25 nm Virion Is the Likely Cause of Transmissible Spongiform

Encephalopathies', *Journal of Cellular Biochemistry*, vol. 100, p. 897 (2007).

p. 85   'The story, to me, is a hideous replay of the tobacco mosaic virus claim of 1936': Laura Manuelidis, personal communication.

p. 85   Wendell Meredith Stanley returned to the United States and settled in New Jersey: Stanley's story is told in L. Kay, 'W. M. Stanley's Crystallization of the Tobacco Mosaic Virus', *Isis*, vol. 77, p. 450 (1986).

p. 86   In 1935 he published a landmark paper: W. Stanley, 'Isolation of a Crystalline Protein Possessing the Properties of Tobacco-Mosaic Virus', *Science*, vol. 81, p. 644 (1935).

p. 87   Bawden and Pirie published their findings in *Nature*: F. Bawden et al., 'Liquid Crystalline Substances from Virus-infected Plants', *Nature*, vol. 138, p. 1051 (1936).

p. 87   explicitly states that the prize was for demonstrating that a virus 'actually is a protein': http://nobelprize.org/nobel_prizes/chemistry/laureates/1946/press.html

p. 87   James Watson mentions Bawden and Pirie's gift explicitly: J. Watson, *The Double Helix* (Weidenfeld & Nicolson, 1968), p. 119.

p. 88   They defined a virus as a piece of nucleic acid that carried genetic information: F. Crick and J. Watson, 'Structure of Small Viruses', *Nature*, vol. 177, p. 4506 (1956).

p. 88   The paper is called 'The "Undiscovered" Discovery': W. Stanley, 'The "Undiscovered" Discovery', *Archives of Environmental Health*, vol. 21, p. 256 (1970).

p. 90   Prusiner has been unwilling to talk to journalists since 1986: Rhodes mentions this in his 'Pathological Science' article in *New Yorker*. Prusiner has talked to a few journalists here and there, but is still reluctant. He declined to be interviewed for the purposes of this book.

p. 90   *Discover* published an article that was highly critical of his methodology: G. Taubes, 'The Game of the Name is Fame. But is it Science?', *Discover*, vol. 7(12), p. 28 (December 1986), available at http://www.slate.com/id/2096/sidebar/42786/

p. 92   *Physics Letters* published a paper by Murray Gell-Mann: M. Gell-Mann, 'A Schematic Model of Baryons and Mesons', *Physics Letters*, vol. 8, p. 214 (1964).

p. 92   Gell-Mann managed to distance himself from any accountability for their existence: M. Gell-Mann, 'Current Algebra: Quarks and What Else?', in *Proceedings of the XVI International Conference on High Energy Physics*, vol. 4, p. 135 (1972).

p. 92   The particle physicist John Polkinghorne caricatured Gell-Mann's
         equivocation: J. Polkinghorne, *Rochester Roundabout: The Story of
         High Energy Physics* (Longman, 1989), p. 110. Before the words I've
         quoted here, Polkinghorne says, 'For many years it was [Gell-Mann's]
         habit to refer to the "presumably mathematical" quark. I always
         considered that to be a coded message.'

p. 93   Carol Reeves ... has carried out a study of Prusiner's rhetorical style:
         As well as drawing on the 'I Knew There Was Something Wrong with
         That Paper' paper mentioned above, this section also draws on the
         fascinating analysis in C. Reeves, 'An Orthodox Heresy: Scientific
         Rhetoric and the Science of Prions', *Science Communication*, vol. 24,
         p. 98 (2002).

p. 94   before they can be 'firmly classified as prions': S. Prusiner, 'Prions:
         Novel Infectious Pathogens', *Advances in Virus Research*, vol. 29, p. 1
         (1984).

p. 94   virologist Richard Carp wrote in 1985: R.I. Carp et al., 'Nature of the
         Scrapie Agent: Current Status of Facts and Hypotheses', *Journal of
         General Virology*, vol. 66, p. 1357 (1985).

p. 95   Carp says that the prion idea is so firmly established: Telephone
         interview with Reeves, cited in 'An Orthodox Heresy'.

p. 95   Scientists are highly resistant to new scientific ideas: Numerous
         historical examples can be found in B. Barber, 'Resistance by
         Scientists to Scientific Discovery', *Science*, vol. 134, p. 596 (1961).

p. 95   'New ideas need the more time for gaining general assent ...': H.
         Helmholtz, 'The Modern Development of Faraday's Conception
         of Electricity', Faraday Lecture, delivered before the Fellows of the
         Chemical Society, London, 5 April 1881. Text available at http://www.
         chemteam.info/Chem-History/Helmholtz-1881.html

p. 95   'none of my professors ... had any understanding for its contents':
         M. Planck, *Scientific Autobiography*, translated by F. Gaynor
         (Philosophical Library, 1968), p. 19.

p. 96   kept science in China and the Far East from leading the world: M.
         Brooks, 'The Spark Rises in the East', *New Statesman*, 16 August
         2010, available at http://www.newstatesman.com/asia/2010/08/
         china-research-chinese-science

p. 96   Today, an estimated 35 million people worldwide have dementia:
         figures available at http://www.alz.co.uk/media/nr100921.html

p. 97   knock out the ability of mice to make prion proteins: C. Le Pichon,
         'Olfactory Behavior and Physiology Are Disrupted in Prion Protein
         Knockout Mice', *Nature Neuroscience*, vol. 12, p. 60 (2008).

p. 97   may not generate new neurons at quite the normal rate: C.C. Zhang et al., 'Prion Protein Is Expressed on Long-Term Repopulating Hematopoietic Stem Cells and Is Important for Their Self-Renewal', *Proceedings of the National Academy of Sciences*, vol. 103, p. 2184 (2006).

p. 97   the absence of prion proteins seems to make long-term memory more robust: E. Parkin et al., 'Cellular Prion Protein Regulates β-Secretase Cleavage of the Alzheimer's Amyloid Precursor Protein', *Proceedings of the National Academy of Sciences*, vol. 104, p. 11062 (2007). It is worth noting that this paper was edited by Stanley Prusiner prior to publication: http://www.pnas.org/content/104/26/11062. The questionable scientific ethics behind some papers in the *Proceedings* are briefly discussed in Chapter 6.

p. 97   When Alois Alzheimer first presented his new disease: N. Zilka and M. Novak, 'The Tangled Story of Alois Alzheimer', *Bratisl Lek Listy*, vol. 107, p. 343 (2006), available at http://www.bmj.sk/2006/107910-02.pdf

p. 97   In 2009, researchers were excited to discover: H. Ledford, '"Harmless" Prion Protein Linked to Alzheimer's Disease', *Nature News Service*, 25 February 2009, http://www.nature.com/news/2009/090225/full/news.2009.121.html

p. 97   human brain cells engineered to make more prion proteins: E. Parkin et al., 'Cellular Prion Protein Regulates β-Secretase Cleavage of the Alzheimer's Amyloid Precursor Protein'.

p. 98   Misfolding is not uncommon in proteins: For a useful essay on protein folding, see W.A. Thomasson, 'Unraveling the Mystery of Protein Folding', http://iop.vast.ac.vn/theor/conferences/smp/1st/kaminuma/UnravelingtheMysteryofProteinFolding/protein.html

p. 98   they involve prion proteins that have somehow folded up differently to normal: A. Aguzzi, 'Beyond the Prion Principle', *Nature*, vol. 459, p. 924 (2009).

p. 99   Normal mice remained perfectly healthy: This experiment is discussed in J. Collinge, 'A General Model of Prion Strains and Their Pathogenicity', *Science*, vol. 318, p. 930 (2007).

p. 99   There is certainly still room for the involvement of a nucleic acid element: For a contrarian view on Prusiner, see L. Maneulidis, 'Transmissible Encephalopathy Agents: Virulence, Geography and Clockwork', *Virulence*, vol. 1, p. 1 (2010).

CHAPTER 4

p. 101   eight young men who had each been paid £2,330: there were global
press reports at the time. For a later analysis, see e.g. D. Leppard,
'Elephant Man Drug Victims Told to Expect Early Death', *Times*, 30
July 2006, available at http://www.timesonline.co.uk/tol/news/uk/
article694634.ece; and S. Hattenstone, "'Everybody Thought We Were
Toxic Waste'", *Guardian*, 17 February 2007, available at http://www.
guardian.co.uk/society/2007/feb/17/health.lifeandhealth. A digest is
available at http://www.i-sis.org.uk/LDTC.php

p. 102   things 'could have been done better': M. Goodyear, 'Further Lessons
from the TGN1412 Tragedy', *British Medical Journal*, vol. 333, p. 270
(2006).

p. 103   In 2005, ethicists Patricia Keith-Spiegel and Gerald Koocher
published a rather enlightening paper: P. Keith-Spiegel and G.
Koocher, 'The IRB Paradox: Could the Protectors Also Encourage
Deceit?', *Ethics and Behavior*, vol. 15, p. 339 (2005), available at www.
ethicsresearch.com/images/IRB_Paradox_EandB.pdf. See also J.
Giles, 'Researchers Break the Rules in Frustration at Review Boards',
Nature, vol. 438, p. 136 (2005); A. Dove, 'Further Concern over Rules
That Impede Research', *Nature Medicine*, vol. 8, p. 5 (2002).

p. 104   'Who knows how many beneficial drugs are being withheld from the
public': Altman raises the question in L. Altman, *Who Goes First?
The Story of Self-Experimentation in Medicine* (Random House 1986).
Altman is the best source for many of the stories in this chapter, in
particular the story of Werner Forssmann.

p. 106   The picture was published in Forssmannn's breakthrough paper:
W. Forssmann, 'Die Sondierung des rechten Herzens', *Klinische
Wochenschrift*, vol. 45, p. 2085 (1929).

p. 107   eventually became Hitler's Surgeon General to the Army: Marc
Dewey and colleagues suggest that Sauerbruch was not all bad:
he 'supported victims of Nazi persecution, attempted to use his
influence to put a stop to the "Euthanasia Program T4", and in private
expressed his criticism of National Socialists'. M. Dewey et al., 'Ernst
Ferdinand Sauerbruch and His Ambiguous Role in the Period of
National Socialism', *Annals of Surgery*, vol. 244, p. 315 (2006).

p. 107   the Allies had experimented on their own citizens and
soldiers: See, for example, P. Cockburn, 'US Navy Tested
Mustard Gas on Its Own Sailors', *Independent*, 14 March 1993,
available at http://www.independent.co.uk/news/world/

us-navy-tested-mustard-gas-on-its-own-sailors-in-1943-the-
americans-used-humans-in-secret-experiments-patrick-
cockburn-in-washington-reports-on-the-survivors-who-bear-the-
scars-1497508.html; and R. Evans, 'Military Scientists Tested Mustard
Gas on Indians', *Guardian*, 1 September 2007, available at http://www.
guardian.co.uk/uk/2007/sep/01/india.military

p. 107   And so arose the defining ethical guidelines in medical science: The
Nuremberg Code can be viewed at http://ohsr.od.nih.gov/guidelines/
nuremberg.html

p. 108   Not that all wartime science was mired in dark deeds: Haldane's
exploits are detailed in R. Clark, *J. B. S.: The Life and Work of J.
B. S. Haldane* (Quality Book Club, 1968), available at http://www.
gyanpedia.in/Portals/0/Toys%20from%20Trash/Resources/books/
haldanebio.pdf

p. 109   the three surgeons were jointly presented with the Nobel Prize
in Physiology or Medicine: http://nobelprize.org/nobel_prizes/
medicine/laureates/1956/forssmann-bio.html

p. 109   a moment of eye-watering recklessness: Newton's illustrated account
of his experiment is available at http://www.newtonproject.sussex.
ac.uk/view/extract/normalized/NATP00004/start=par62&end=par64

p. 110   The nineteenth-century pioneers of anaesthesia: Altman's *Who Goes
First* is detailed on the history of anaesthesia, but you could also visit
http://www.general-anaesthesia.com/#historical

p. 111   He was, in his own words, 'a little scientist guy': The quotes come
from a fascinating and provocative TED (Technology, Entertainment,
Design) talk available at http://www.ted.com/index.php/talks/
kary_mullis_on_what_scientists_do.html

p. 111   Sources for Barry Marshall's story: 'Helicobacter Connections', Nobel
Lecture, 8 December 2005, available at http://nobelprize.org/nobel_
prizes/medicine/laureates/2005/marshall-lecture.html; B. Marshall
(editor), *Helicobacter Pioneers: Firsthand Accounts from the Scientists
Who Discovered Helicobacters 1892–1982* (Blackwell, 2002).

p. 118   Warren actually has no memory of the incident: Robin Warren,
personal communication.

p. 121   Daniel Boorstin: 'The greatest obstacle to knowledge is not ignorance;
it is the illusion of knowledge': This quote comes from an interview
with Boorstin: C. Krucoff, 'The 6 O'Clock Scholar', *Washington Post*,
29 January 1984, p. K1.

p. 125   the first line of attack for doctors presented with a peptic ulcer: http://
consensus.nih.gov/1994/1994HelicobacterPyloriUlcer094html.htm

p. 126   Major Walter Reed, left Cuba and headed back to Washington: An interesting, up-to-date take on the incident can be found in A. Mehra, 'Politics of Participation: Walter Reed's Yellow-Fever Experiments', *Virtual Mentor*, vol. 11, p. 326 (April 2009), available at http://virtualmentor.ama-assn.org/2009/04/mhst1–0904.html

p. 126   the shameful US Public Health Service project known as the Tuskegee syphilis study: There are many books on this episode, e.g. S. Reverby, *Examining Tuskegee: The Infamous Syphilis Study and Its Legacy* (University of North Carolina Press, 2009), but there are also a host of online resources about it, e.g. the website of the US Centers for Disease Control and Prevention, http://www.cdc.gov/tuskegee/index.html

p. 126   An equally shameful but more recent example of scientific recklessness: The best commentary and rundown of the Wakefield scandal is at the website of *Sunday Times* journalist Brian Deer: http://briandeer.com/mmr-lancet.htm

p. 127   'a callous disregard' for the suffering of children: The GMC statement, issued on 24 May 2010, is available at www.gmc-uk.org/Wakefield_SPM_and_SANCTION.pdf_32595267.pdf

p. 127   Measles has again emerged as a killer disease: There is a salutary warning of Wakefield's impact in the UK Health Protection Agency's graph of notifications and confirmed cases at http://www.hpa.org.uk/web/HPAweb&HPAwebStandard/HPAweb_C/1195733808276

p. 127   Hwang is not guilt-free: See, for example, 'S Korea cloning pioneer disgraced', 24 November 2005, http://news.bbc.co.uk/1/hi/world/asia-pacific/4465552.stm; D. Cyranoski, 'Clone Star Admits Lies over Eggs', *Nature*, vol. 438, p. 536 (2005); '10 Questions For Dr. Hwang Woo Suk', http://www.time.com/time/magazine/article/0,9171,1137709,00.html; 'Cloning Pioneer Admits Ethical Violations and Quits', http://www.newscientist.com/article/dn8367

p. 127   There are plenty more examples of questionable practices: See e.g. J. Minkel, 'Self-Experimenters Step Up for Science', scientificamerican.com, 10 March 2008, available at http://www.scientificamerican.com/article.cfm?id=self-experimenters

p. 128   As Rebecca Skloot observes: R. Skloot, *The Immortal Life of Henrietta Lacks* (Macmillan, 2010), p. 315.

p. 129   the 'indescribable' feeling of having worms burrow through his skin: E. Svoboda, 'The Worms Crawl in', *New York Times*, 1 July 2008, available at http://www.nytimes.com/2008/07/01/health/research/01prof.html

p. 130   an unlikely sentence in *The Biochemist*: D. Pritchard, 'Can Parasites Be Good for You?', *The Biochemist*, vol. 31(4), p. 28 (2009).

p. 132   'I would have died in the service of science ...': from E.E. Hume, *Max von Pettenkofer: His Theory of the Etiology of Cholera, Typhoid Fever, and Other Intestinal Diseases. A Review of His Arguments and Evidence* (Hoeber, 1927), p. 456.

CHAPTER 5

p. 133   Landrum Shettles was hurrying towards his laboratory: The story is told in PBS's American Experience programme *Test Tube Babies*. There is a transcript at http://www.pbs.org/wgbh/americanexperience/features/transcript/babies-transcript/

p. 135   the EAB finally issued their report on the prospects for human IVF: For a discussion of the report and its context, see http://www.nap.edu/openbook.php?record_id=11278&page=23. This is part of *Guidelines for Human Embryonic Stem Cell Research* (National Academies Press, 2005). See also this useful timeline on the history and politics of embryo and stem cell research in the United States: http://www.nature.com/gt/journal/v9/n11/full/3301744a.html

p. 135   Americans now favoured the use of IVF: For the polling history, see H. Mason Kiefer, 'Gallup Brain: The Birth of In Vitro Fertilization', 5 August 2003, available at http://www.gallup.com/poll/8983/gallup-brain-birth-vitro-fertilization.aspx

p. 136   'What used to be something that most of our ancestors thought ...': The quote comes from commentaries accompanying the American Experience *Test Tube Babies* documentary, available at http://www.pbs.org/wgbh/americanexperience/features/interview/ethical-questions/

p. 136   Robert Edwards and his wife had made friends with a couple who were childless: R. Edwards, *A Matter of Life* (Morrow, 1980), p. 38.

p. 136   *Human Reproduction* published an extraordinary article: M. Johnson et al., 'Why the Medical Research Council Refused Robert Edwards and Patrick Steptoe Support for Research on Human Conception in 1971', *Human Reproduction*, vol. 25, p. 2157 (2010).

p. 137   Around four million people have been conceived through IVF: C. Russell, 'Four Million Test-Tube Babies and Counting', *The Atlantic*, 7 October 2010, available at http://www.theatlantic.com/technology/archive/2010/10/four-million-test-tube-babies-and-counting/64198/

p. 138 When Edwards won his Nobel Prize: http://nobelprize.org/nobel_prizes/medicine/laureates/2010/press.html

p. 138 'In retrospect, it is fortunate that Edwards and Steptoe pressed on': J. Biggers, 'Editorial', *Human Reproduction*, vol. 25, p. 2156 (2010), available at http://humrep.oxfordjournals.org/content/early/2010/07/22/humrep.deq156.full.pdf

p. 139 it takes forty days for an embryo to form in the mother's womb: E.N. Dorff, 'Catholics, Jews and Petri Dishes', *Forward*, 13 October 2010, http://www.forward.com/articles/132119/

p. 139 'If the embryo is still unformed ...': The quote comes from Augustine's *On Exodus*, 21.80. It is quoted in B. Rowland, *Medieval Woman's Guide to Health: The First English Gynecological Handbook* (Kent State University Press, 1981), p. 36.

p. 140 anyone associated with the process of abortion: See http://faculty.cua.edu/Pennington/Law111/CatholicHistory.htm

p. 140 the process had become routinely successful in dogs, horses, foxes and rabbits: R. Foote, 'The History of Artificial Insemination: Selected Notes and Notables', *Journal of Animal Science*, vol. 80, p. 1 (2002), available at www.asas.org/symposia/esupp2/Footehist.pdf

p. 141 there is an 'inseparable connection ...': Pope Paul VI's encyclical *Humanae Vitae* is available at http://www.vatican.va/holy_father/paul_vi/encyclicals/documents/hf_p-vi_enc_25071968_humanae-vitae_en.html

p. 141 It has since banned its members from using IVF: Catechism 2377 says: 'Techniques involving only the married couple (homologous artificial insemination and fertilization) are perhaps less reprehensible, yet remain morally unacceptable.' See http://www.vatican.va/archive/ccc_css/archive/catechism/p3s2c2a6.htm

p. 141 'an infinitely precious human service': P. Lewis, 'Catholic Hospitals in Europe Defy Vatican on In-Vitro Fertilization', *New York Times*, 18 March 1987, available at http://www.nytimes.com/1987/03/18/world/catholic-hospitals-in-europe-defy-vatican-on-in-vitro-fertilization.html

p. 141 no one paid such pronouncements any attention: N.D. Kristof, 'Australians Cool to Vatican Paper', *New York Times*, 19 March 1987, available at http://www.nytimes.com/1987/03/19/world/australians-cool-to-vatican-paper.html?src=pm

p. 141 68 per cent of American Catholics approved of artificial contraception: A.L. Greil, 'The Religious Response to Reproductive

Technology', *Christian Century*, 4 January 1989, available at http://
www.religion-online.org/showarticle.asp?title=807

p. 141 Catholics also strongly support IVF: P. Lauritzen, 'Catholics &
IVF: The Next Big Battleground?', *Commonweal*, 12 August 2005,
available at http://findarticles.com/p/articles/mi_m1252/is_14_132/
ai_n27859685/

p. 142 'no one can be the arbiter of life except God himself': This quote is
actually from Pope John Paul II, who was in turn quoting from Pope
Paul VI's 1968 *Humanae Vitae* and Pope John XXIII's 1961 encyclical
Mater et Magistra.

p. 142 in his autobiography, *A Life Decoded*: J. Venter, *A Life Decoded: My
Genome: My Life* (Allen Lane, 2007). This is the main source for this
section and its quotes.

p. 143 in James Shreeve's *The Genome War*: J. Shreeve, *The Genome War*
(Fawcett Books, 2005).

p. 143 Einstein expressed the scientist's view with a typical simplicity: A.
Einstein, 'Letter to an Admirer', 24 March 1951, quoted in H. Dukas
and B. Hoffmann (editors), *Albert Einstein: The Human Side. New
Glimpses from His Archives* (Princeton University Press, 1979), p. 57.

p. 143 Many biologists working in the public sector rejoiced: J. Kaiser,
'Celera to End Subscriptions and Give Data to Public GenBank',
Science, vol. 308, p. 775 (2005).

p. 144 Even Venter's competitors and detractors were largely unwilling to
score points: N. Wade, 'Scientist Reveals Secret of Genome: It's His',
New York Times, 27 April 2002, available at http://www.nytimes.
com/2002/04/27/us/scientist-reveals-secret-of-genome-it-s-his.html

p. 144 'This is the first self-replicating species we've had on the planet whose
parent is a computer': A recording of the press conference is available
at http://www.ted.com/talks/craig_venter_unveils_synthetic_life.html

p. 145 Venter's science raised 'genuine concerns', Barack Obama said:
There is a good summary with links to other sources at http://www.
biopoliticaltimes.org/article.php?id=5220

p. 146 'He has not created life, only mimicked it': N. Wade, 'Researchers
Say They Created a "Synthetic Cell"', *New York Times*, 20 May 2010,
available at http://www.nytimes.com/2010/05/21/science/21cell.html

p. 146 'a defining moment in the history of biology and biotechnology': I.
Sample, 'Craig Venter Creates Synthetic Life Form', *Guardian*, 20 May
2010, available at http://www.guardian.co.uk/science/2010/may/20/
craig-venter-synthetic-life-form

p. 146 'great need for more oversight of this hugely powerful technology': A. Caplan, 'Now Ain't That Special? The Implications of Creating the First Synthetic Bacteria', *Scientific American* guest blog, 20 May 2010, http://www.scientificamerican.com/blog/post. cfm?id=now-aint-that-special-the-implicati-2010–05–20

p. 146 Venter's press conference and publication in *Science*: D. Gibson et al., 'Creation of a Bacterial Cell Controlled by a Chemically Synthesized Genome', *Science*, vol. 329, p. 52 (2010).

p. 146 'the implications of this scientific milestone ...': D. Brown, 'Scientists Create Cell Based on Man-Made Genetic Instructions', *Washington Post*, 21 May 2010, available at http://www.washingtonpost.com/ wp-dyn/content/article/2010/05/20/AR2010052003336.html

p. 147 'The stuff that's coming down the pipe ...': T. Reichhardt, 'Studies of Faith', *Nature*, vol. 432, p. 666 (2004).

p. 148 The research first came to the attention of the world's press: K. Nayernia, 'In Vitro-Differentiated Embryonic Stem Cells Give Rise to Male Gametes that Can Generate Offspring Mice', *Developmental Cell*, vol. 11, p. 125 (2006).

p. 149 sperm from adult human stem cells: K. Nayernia, 'Derivation of Male Germ Cells from Bone Marrow Stem Cells', *Laboratory Investigation*, vol. 86, p. 654 (2006).

p. 150 The eggs showed some signs of attempting cell division: K. Hübner et al., 'Derivation of Oocytes from Mouse Embryonic Stem Cells', *Science*, vol. 300, p. 1251 (2003).

p. 150 grown cells from testes in a liquid broth: O. Lacham-Kaplan, 'Testicular Cell Conditioned Medium Supports Differentiation of Embryonic Stem Cells into Ovarian Structures Containing Oocytes', *Embryonic Stem Cells*, vol. 24, p. 266 (2006).

p. 151 added bone proteins to her stem cell mix: K. Keel et al., 'Human DAZL, DAZ and BOULE Genes Modulate Primordial Germ-Cell and Haploid Gamete Formation', *Nature*, vol. 462, p. 222 (2009).

p. 151 she made capsules of gel from a chemical found in seaweed: E. West et al., 'Engineering the Follicle Microenvironment', *Seminars in Reproductive Medicine*, vol. 25, p. 287 (2007).

p. 151 In 2009, her team managed to incubate human follicles: If you want the latest, it's worth a visit to http://www.woodrufflab.org/

p. 151 The procedures involved have 'turned out to be much simpler than we ever dreamed': Quoted from H. Leggett, 'A Fertility First: Human Egg Cells Grow Up in Lab', 14 July 2009, http://www.wired.com/ wiredscience/2009/07/humanegg/

p. 152 building a womb from a few cells taken from the endometria: This is
discussed in detail in G. Reynolds, 'Artificial Wombs: Will We Grow
Babies Outside Their Mothers' Bodies?', *Popular Science*, August
2005, available at http://www.popsci.com/scitech/article/2005-08/
artificial-wombs; and F. Dolendo, 'Baby Machines: The Birth of the
Artificial Womb', *The Triple Helix*, vol. 2(2), p. 4 (2006), available at
http://stuff.mit.edu/afs/athena/activity/t/triplehelix/archive/MIT%20
Final2.pdf

p. 152 'third era of human reproduction': S. Welin, 'Reproductive
Ectogenesis: The Third Era of Human Reproduction and Some Moral
Consequences', *Science and Engineering Ethics*, vol. 10, p. 615 (2004).

p. 153 And it *will* happen, Gosden says: R. Gosden, *Designer Babies: Science
and the Future of Human Reproduction* (Victor Gollancz, 1999), p.
180. See also H. Pearson, 'Making Babies: The Next 30 Years', *Nature*,
vol. 454, p. 260 (2008).

p. 153 it might take sixty years, he suggests, but it is 'inevitable': The quote
comes from P. Klass, 'The Artificial Womb Is Born', *New York Times
Magazine*, 29 September 1996, available at http://www.nytimes.
com/1996/09/29/magazine/the-artificial-womb-is-born.html

p. 154 Jeremy Rifkin issued an apocalyptic warning: J. Rifkin and E.
Howard, *Who Should Play God?* (Delacorte Press, 1977), p. 115.

p. 154 Rifkin has since suggested that children nurtured in an artificial
womb: J. Rifkin, 'The End of Pregnancy', *Guardian*, 17 January 2002,
available at http://www.guardian.co.uk/world/2002/jan/17/gender.
medicalscience

p. 155 Huxley declared that he would write it differently: See http://
en.wikipedia.org/wiki/Island_(novel)

p. 156 As far back as 1971, Edward Grossman ... pointed out: Quoted in A.
Alghrani, 'The Legal and Ethical Ramifications of Ectogenesis', *Asian
Journal of WTO & International Health Law and Policy*, vol. 2, p. 189
(2007).

p. 156 'I find ectogenesis in many ways repugnant': S. Welin, 'Reproductive
Ectogenesis: The Third Era of Human Reproduction and Some Moral
Consequences'.

p. 156 As Roger Gosden has recently said: R. Gosden, 'Roger Gosden
Forecasts the Future', *New Scientist*, 21 November 2006, available
at http://www.newscientist.com/article/dn10625-roger-gosden-
forecasts-the-future.html

p. 157 fewer than twenty of our 20,000 or so genes are unique to humans: D. Knowles and A. McLysaght, 'Recent de novo Origin of Human Protein-Coding Genes', *Genome Research*, vol. 19, p. 1752 (2009).

p. 158 *The Great Ape Project* demanded: P. Singer (editor), *The Great Ape Project* (Fourth Estate, 1993).

p. 158 Fernando Sebastián, the archbishop of Pamplona and Tudela, called the idea ridiculous: 'Spanish MPs Push for Apes' Rights', 8 June 2006, http://news.bbc.co.uk/1/hi/world/europe/5058986.stm

p. 159 Copernicus's source was as strange and irrational as Einstein's: P. Feyerabend, *Against Method*, Verso (1993), p. 39. Feyerabend also discusses the reactions of Galileo and Ptolemy.

p. 161 Stephen Hawking had declared, Laplace-like, that God is defunct as Creator: H. Devlin, 'Hawking: God Did Not Create Universe', *Times*, 2 Sept 2010.

p. 161 Hawking tells the story of a 1981 conference on cosmology: S. Hawking, *A Brief History of Time* (Bantam, 1995), p. 116.

CHAPTER 6

p. 163 'I played over the music of that scoundrel Brahms ...': These examples are drawn from C. Cerf and V. Navasky, *The Experts Speak: The Definitive Compendium of Authoritative Misinformation* (Villard, 1998), via an eye-opening essay by Thomas Gold: 'New Ideas in Science', *Journal of Scientific Exploration*, vol. 3, p. 103 (1989).

p. 164 Lewis was found dead in his laboratory: The lives and rivalries of Lewis, Langmuir and Arrhenius can be explored in P. Coffey, *Cathedrals of Science* (Oxford University Press, 2008).

p. 164 his celebrated 'Pathological Science' essay: I. Langmuir, 'Pathological Science', unpublished but available at http://www.cs.princeton. edu/~ken/Langmuir/langmuir.htm

p. 165 a study by Dutch researchers found that schoolchildren who were timid and introverted: J. Marchant, 'Should Schoolchildren Be Typecast into Science?' *New Scientist*, 22 October 2010, p. 14.

p. 165 'Much of a scientist's pride and sense of accomplishment': P. Medawar, 'The Act of Creation', *The Strange Case of the Spotted Mice and Other Classic Essays on Science* (Oxford University Press, 1996), p. 41 (first published in *New Statesman*, 19 June 1964).

p. 165   Take the story of the transistor, for example: For the full tale, only sketched in this section, read M. Riordan and L. Hoddeson, *Crystal Fire: The Birth of the Information Age* (Norton, 1997).

p. 169   'He set extraordinarily high standards for everyone, including himself …': 'Memorial Resolution: William B. Shockley (1910–1989)', available at http://histsoc.stanford.edu/pdfmem/ShockleyW.pdf

p. 169   Isaac Newton and Friedrich Gauss, who both waited twenty years for recognition and acceptance: D. Watson, *Scientists Are Human* (Arno Press, 1975), p. 57.

p. 169   'When a true genius appears in this world …': J. Swift, 'Essay on the Fates of Clergymen' (1728).

p. 170   'A conviction of the fundamental soundness of the idea took root in my mind': http://www.ucmp.berkeley.edu/history/wegener.html

p. 170   Sir John Lubock described it as 'nothing but nonsense': D. Ebbing and S. Gammon, *General Chemistry* (Houghton Mifflin, 2008), p. 199.

p. 171   'just an old bag who'd been hanging around Cold Spring Harbor for years': Barbara McClintock's difficult journey is chronicled in E. Fox Keller, *A Feeling for the Organism* (Freeman, 1983).

p. 171   MRSA, for instance, contributed to the death of around 13,000 Britons between 1993 and 2009: UK Office for National Statistics, details available at http://www.statistics.gov.uk/cci/nugget. asp?id=1067

p. 174   The science writer Matt Ridley has an interesting way to describe chromosomes: M. Ridley, *Genome* (Harper Perennial, 2004), p. 6.

p. 174   by 1951 McClintock was able to publish a stab at what was going on: B. McClintock, 'Chromosome Organization and Genic Expression,' *Cold Spring Harbor Symposia on Quantitative Biology*, vol. 16, p. 40 (1951).

p. 175   'This became painfully evident to me …': The National Library of Medicine has collected McClintock's letters and papers at http:// profiles.nlm.nih.gov/LL/

p. 176   'Research uses real egotists who seek their own pleasure …': A. Szent-Györgyi, *Science Today*, May 1980, p. 35.

p. 176   'I just go my own pace here': quoted in L. Kass, 'Records and Recollections: A New Look at Barbara McClintock, Nobel-Prize-Winning Geneticist', *Genetics*, vol. 164, p. 1251 (August 2003).

p. 177   the French geneticists François Jacob, André Lwoff and Jacques Monod won their Nobel Prize: http://nobelprize.org/nobel_prizes/ medicine/laureates/1965/

p. 177   'their thinking was probably much influenced by Barbara's notion':
This and the following quote are from N. Comfort, 'From Controlling
Elements to Transposons: Barbara McClintock and the Nobel Prize',
Trends in Genetics, vol. 17, p. 475 (2001).

p. 177   The first drafts of her acceptance speech: These are in the National
Library of Medicine collection of McClintock's papers (see Note for p.
175).

p. 178   'I was not invited to give lectures or seminars ...': This quote is from
McClintock's Banquet Speech, available at http://nobelprize.org/
nobel_prizes/medicine/laureates/1983/mcclintock-speech.html

p. 179   the sentiment expressed by the French physiologist Claude Bernard:
C. Bernard et al., *Introduction to Experimental Medicine* (Dover
Publications, 1957), p. 227.

p. 180   In July 2010, Nobel manoeuvrings came to the fore: Z. Merali,
'Physicists Get Political over Higgs', *Nature News*, http://www.nature.
com/news/2010/100804/full/news.2010.390.html

p. 180   The problem is one of priority: Ian Sample, author of *Massive: The
Hunt for the God Particle* (Virgin Books, 2010), has laid out the issues
and the chronology at http://www.iansample.com/site/?q=content/
higgs-row-and-nobel-reform

p. 181   'Anyone who witnesses the advance of science first-hand ...': C.
Sagan, *The Demon-Haunted World* (Ballantine, 1996), p. 255.

p. 182   that much is clear from reading what other scientists say about it:
These commentaries, including Margulis's own, come from 'Gaia
Is a Tough Bitch' in J. Brockman (editor), *The Third Culture* (Simon
& Schuster, 1995), Chapter 7, available at http://www.edge.org/
documents/ThirdCulture/n-Ch.7.html

p. 182   Andreas Schimper made an interesting observation: Margulis tells
the story of endosymbiotic theory in L. Margulis and D. Sagan,
Acquiring Genomes: A Theory of the Origins of Species (Basic Books,
2002). A more thorough scientific account is in L. Margulis, *Origin
of Eukaryotic Cells: Evidence and Research Implications for a Theory*
(Yale University Press, 1970).

p. 183   biological complexity might have arisen from such arrangements
being made permanent: One of Margulis's early attempts to interest
biologists in this history (while she was still married to Carl Sagan)
can be found in L. Sagan, 'On the Origin of Mitosing Cells', *Journal of
Theoretical Biology*, vol. 14, p. 225 (1967).

p. 184   'It may have started when one sort of squirming bacterium invaded
another ...': L. Margulis, 'Gaia is a Tough Bitch'.

p. 185   'I was told by an NSF grants officer …': L. Margulis, 'Peer Review
          Attacked' (letter), *The Sciences*, vol. 17, p. 31 (1977). It is cited in the
          eye-opening B. Martin, 'Bias in Awarding Research Grants', *British
          Medical Journal*, vol. 293, p. 550 (1986), available at http://www.
          bmartin.cc/pubs/86bmj.html

p. 186   Margulis quotes the biologist Carl Clarence Lindegren: The quote
          comes from C. Lindegren, *Cold War in Biology* (Planarian Press,
          1966), p. 10.

p. 186   the fossil record gives us millions of years of data: L. Margulis, 'Gaia
          is a Tough Bitch'.

p. 187   she pulled the prestigious US National Academy of Sciences
          through the dirt: See, for example, B. Borrell, 'National Academy
          as *National Enquirer*? *PNAS* Publishes Theory That Caterpillars
          Originated from Interspecies Sex', *Scientific American* Online, 24
          August 2009, available at http://www.scientificamerican.com/
          article.cfm?id=national-academy-as-national-enquirer. See also:
          Z. Corbyn, 'Probe Leaves Butterfly Paper's Fate Up in the Air',
          Times Higher Education, 1 October 2009, available at http://www.
          timeshighereducation.co.uk/story.asp?sectioncode=26&storycod
          e=408496&c=2; S. Kean, 'Controversy "Proceeding" at National
          Academy's Journal', *Science Insider*, 1 October 2009, available at
          http://news.sciencemag.org/scienceinsider/2009/10/controversy-pro.
          html

p. 187   Donald Williamson is an Eeyore amongst scientists: Margulis
          describes Williamson's ideas in L. Margulis and D. Sagan, *Acquiring
          Genomes: A Theory of the Origins of Species* (Basic Books, 2002),
          p. 165. She describes her initial contact with him in the BBC radio
          documentary 'A Life with … Microbes', http://www.bbc.co.uk/
          programmes/b001k12y. Film-maker Robert Sternberg's fascinating
          PhD thesis includes transcripts of parts of this programme, including
          Williamson's more pessimistic pronouncements: R.J. Sternberg,
          'Discovery as Invention: a Constructivist Alternative to the Classic
          Science Documentary', available at westminsterresearch.wmin.
          ac.uk/8884/1/Robert_STERNBERG.pdf

p. 188   That's how Williamson managed to publish: D. Williamson,
          'Caterpillars Evolved from Onychophorans by Hybridogenesis',
          Proceedings of the National Academy of Sciences, vol. 106, p. 19901
          (2009).

p. 190 'After the very first talk I ever gave at an international symposium ...':
J. Rohn, 'Peer Review Is No Picnic', *Guardian*, 6 Sept 2010, available at
http://www.guardian.co.uk/science/blog/2010/sep/06/peer-review

p. 191 'One likes to think of science as divorced from personalities ...': The
quote comes from C. Lindegren, *Cold War in Biology* (Planarian
Press, 1966).

CHAPTER 7

p. 193 In Chandra's case, the establishment figure was the astronomer
Arthur Eddington: Sources on Chandra and Eddington: A.
Miller, *Empire of the Stars* (Little Brown, 2005); 'Subrahmanyan
Chandrasekhar', a compendium of articles on Chandra compiled
by Andrew Mylwaganam, available at http://www.tamil.net/people/
andrew/subra.htm; E.N. Parker, *Subrahmanyan Chandrasekhar
(1910–1995): A Biographical Memoir* (National Academies Press, 1997),
available at www.nap.edu/html/biomems/schandrasekhar.pdf; E.N.
Parker, 'Subrahmanyan Chandrasekhar', Obituaries, *Physics Today*,
November 1995, p. 106; S. Weart, 'Oral History Transcript – Dr. S.
Chandrasekhar', 31 October 1977, available at http://www.aip.org/
history/ohilist/4551_3.html

p. 196 'The paper which has just been presented is all wrong': The
meeting report is in *The Observatory*, vol. 58, p. 33 (February 1935);
Eddington's attack on Chandra is on p. 37. Both can be accessed at
http://adsabs.harvard.edu/abs/1935obs....58...33

p. 200 along came the Swedish physicist Hannes Alfvén: Sources on
Alfvén: H. Alfvén, 'Plasma Physics, Space Research and the Origin
of the Solar System', Nobel Lecture, 11 December 1970, available at
nobelprize.org/nobel_prizes/physics/laureates/1970/alfven-lecture.
pdf; A. Dessler, 'Nobel Prizes: 1970 Awards Honor Three in Physics
and Chemistry', *Science*, vol. 170, p. 604 (1970); H. Alfvén, 'Memoirs
of a Dissident Scientist', *American Scientist*, vol. 76, p. 249 (May/
June 1998); C.-G. Fälthammar and A. Dessler, 'Hannes Alfvén:
Biographical Memoir', *Proceedings of the American Philosophical
Society*, vol. 150, p. 649 (2006); A.L. Peratt, 'Hannes Alfvén: Dean of
the Plasma Dissidents', *The World & I*, p. 190 (May 1988), available at
http://public.lanl.gov/alp/plasma/people/alfven.html; B. De, 'Hannes

Alfvén Birth Centennial: A Pictorial Tribute', http://www.bibhasde.com/Alfven100.html

p. 201   'It would be difficult to overestimate the great influence which Chapman exerted ...': 'Professor Sydney Chapman: An Outstanding Mathematical Physicist', Obituary, *Times*, 18 June 1970, p. 12, available at http://www-history.mcs.st-andrews.ac.uk/Obits/Chapman.html

p. 201   'A new scientific truth does not triumph by convincing its opponents ...': M. Planck, *Scientific Autobiography and Other Papers*, translated by F. Gaynor (New York, 1949), p. 33.

p. 202   the Norwegian physicist Kristian Birkeland was found dead in a Tokyo hotel room: Birkeland's life and work is fascinatingly documented in L. Jago, *The Northern Lights* (Penguin, 2002).

p. 202   'It seems to be a natural consequence of our points of view ...': K. Birkeland, 'Polar Magnetic Phenomena and Terrella Experiments', *The Norwegian Aurora Polaris Expedition 1902–1903*, Section 2 (H. Aschehoug & Co., 1913), p. 720, available at http://www.archive.org/details/norwegianaurorapo1chririch

p. 202   it was not until 1963 that Birkeland was proved right: T. Potemra, 'Birkeland Currents, Recent Contributions from Satellite Magnetic Field Measurements', *Physica Scripta*, vol. 18, p. 152 (1987).

p. 203   'I have no trouble publishing in Soviet astrophysical journals ...': H. Alfvén, 'Memoirs of a Dissident Scientist'.

p. 203   In the end, he managed to publish what is now the foundation of our modern understanding: H. Alfvén, 'A Theory of Magnetic Storms and the Aurorae', *Kungliga Svenska Vetenskapsakademiens Handlingar* III, vol. 18, p. 3 (1939).

p. 204   a research report on the dangers posed by the Sun: *Severe Space Weather Events: Understanding Societal and Economic Impacts* (National Academies Press, 2008), available at http://www.nap.edu/catalog.php?record_id=12507

p. 205   In 2010, President Obama gave NASA a new objective: T. Malik, 'Obama Aims to Send Astronauts to an Asteroid, Then to Mars', 15 April 2010, http://www.space.com/8222-obama-aims-send-astronauts-asteroid-mars.html

p. 206   The workers' union took over the railways: G. Bailey, 'Anarchists in the Spanish Civil War', *International Socialist Review*, no. 24, July/August 2002, available at http://www.isreview.org/issues/24/anarchists_spain.shtml

p. 206   Every year, the club donates 0.7 per cent of its income to UNICEF: R. Karwal, 'FC Barcelona Supports UNICEF's Fight Against HIV/AIDS',

UNICEF Newsline, 1 December 2008, available at http://www.unicef.
org/aids/index_46705.html

p. 207   made the front page of the *New York Times*: W. Stevens, 'New,
Glassy Electronic Device May Outstrip the Transistor', *New
York Times*, 11 November 1968, quoted in H. Wasserman,
'From PCs to PV: A Profile of Ovshinsky Electronics',
Photovoltaics World, 17 January 2007, available at http://
www.renewableenergyworld.com/rea/news/article/2007/01/
from-pcs-to-pv-a-profile-of-ovshinsky-electronics-51559

p. 207   Ovshinsky's parents were immigrants from Eastern Europe:
Biographical sources on Ovshinsky include B. Schwartz and H.
Fritzsche, *Stanford R. Ovshinsky: The Science and Technology of an
American Genius* (World Scientific Publishing, 2008), Chapter 1,
available at www.worldscibooks.com/etextbook/6877/6877_chap01.
pdf

p. 210   Ovshinsky's patent has now been licensed by the chip-maker Intel:
'The Edison of Our Age?', *The Economist Technology Quarterly*,
30 November 2006, available at http://www.economist.com/
node/8312367

p. 210   A February 1970 article in the magazine *Science & Mechanics*:
N. Carlisle, 'The Ovshinsky Invention', *Science & Mechanics*,
February 1970, p. 38, available at http://blog.modernmechanix.
com/2009/02/12/the-ovshinsky-invention/

p. 211   has been referred to as 'Japan's American Genius': This was the title of
a 1987 PBS Nova documentary about Ovshinsky's work.

p. 211   *Forbes* magazine has called him 'the inventor who can create anything
but profits': J. Fahey, 'Repeat Pretender', *Forbes*, 24 November 2003, p.
86, available at http://www.forbes.com/forbes/2003/1124/086.html

p. 211   their aim was to use 'creative science to solve societal problems': 'The
Edison of Our Age?'.

p. 212   He was also still a member of their union: 'The Edison of Our Age?'.

p. 212   'A lot of my best ideas came from Stan': A. Bienenstock, 'Bienenstock
on Ovshinsky', *Berkeley Review of Latin American Studies*, Spring
2008, p. 25, available at http://www.clas.berkeley.edu/Publications/
Review/Spring2008/pdf/BRLAS-Spring2008-ovshinsky-bienenstock.
pdf

p. 212   'I'll have to throw a party for fifty people': R. Walgate, 'Nobel Prizes
1977', *New Scientist*, 20 October 1977, p. 146. There is a photo of
Mott pouring from this enormous bottle of champagne in E.A.
Davis (editor), *Nevill Mott: Reminiscences and Appreciations*

(Taylor & Francis, 1998), p. 176e, available at http://www. ebook3000.com/Biographies/Nevill-Mott--Reminiscences-And-Appreciations_104635.html

p. 212   its investors decided that they'd had enough of not turning a profit: D. Buss, 'At 85, Green Giant Stan Ovshinsky Sees His Ideas Bearing Fruit', *Green Car Advisor*, 7 July 2008, available at http://blogs. edmunds.com/greencaradvisor/2008/07/profile-at-85-green-giant-stan-ovshinsky-sees-his-ideas-bearing-fruit.html

p. 212   '... I use the periodic chart of atoms as if it's an engineering diagram': M. Villiger, 'Meet the Ovshinskys', *Scientific American Frontiers*, 19 May 2004, available at http://www.pbs.org/saf/1506/features/ ovshinsky2.htm

p. 213   'Scientists tend to resist interdisciplinary inquiries ...': Quoted in A.L. Peratt, 'Hannes Alfvén: Dean of the Plasma Dissidents'.

p. 213   'From each according to his abilities, to each according to his needs': K. Marx, 'Critique of the Gotha Program' (1875), available at http:// www.marxists.org/archive/marx/works/1875/gotha/ch01.htm

p. 213   as sociologist Robert Merton pointed out: R.K. Merton, 'The Matthew Effect in Science: The Reward and Communication Systems of Science Are Considered', *Science*, vol. 159, p. 56 (1968), available at www.garfield.library.upenn.edu/merton/matthew1.pdf

p. 214   J.B.S. Haldane noticed it at work in his sphere: Merton discussed this incident in his follow-up paper: R. Merton, 'The Matthew Effect in Science II: Cumulative Advantage and the Symbolism of Intellectual Property', *Isis*, vol. 79, p. 606 (1988), available at http://garfield.library. upenn.edu/merton/matthewii.pdf

p. 215   Alfvén spoke with bitterness of his status as a 'dissident': H. Alfvén, 'Memoirs of a Dissident Scientist'.

p. 215   'I'm not a part of their world': 'The Edison of Our Age?'.

CHAPTER 8

p. 216   the atmosphere is anything but relaxed: I attended this meeting, which was given a brief write-up in *The Economist*, available as 'Phoning ET' at: http://www.economist.com/node/17199376

p. 219   a direct consequence of Sagan's scientific studies: R. Turco et al., 'Nuclear Winter: Global Consequences of Multiple Nuclear Explosions', *Science*, vol. 23, p. 1283 (1983).

p. 219   He applied this understanding to the scenario of all-out nuclear war: C. Sagan, 'Nuclear Winter', available at http://www. cooperativeindividualism.org/sagan_nuclear_winter.html

p. 220   denied membership of the US National Academy of Sciences: G. Benford, 'A Tribute to Carl Sagan: Popular & Pilloried', *Skeptic*, vol. 13, no. 2, available at http://www.skeptic.com/reading_room/popular-and-pilloried/

p. 220   Sagan was 'a nobody' who 'never did anything worthwhile.': K. Davidson, *Carl Sagan: A Biography* (Wiley, 1999), p. 380.

p. 221   Michael Shermer took it upon himself: M. Shermer, 'The Measure of a Life', *Skeptic*, vol. 7, no. 4, available at http://www.skeptic.com/reading_room/the-measure-of-a-life/

p. 221   Once released, CFCs are stable in the atmosphere: The discussion of ozone and CFCs is largely derived from S. Roan, *Ozone Crisis: The 15-Year Evolution of a Sudden Global Emergency* (Wiley, 1989); J. Gribbin, *The Hole in the Sky: Man's Threat to the Ozone Layer* (Bantam, 1988); L. Dotto and H. Schiff, *The Ozone War* (Doubleday, 1978); N. Oreskes and E. Conway, *Merchants of Doubt*, (Bloomsbury, 2010).

p. 223   if CFC production had carried on unchecked: World Health Organization, 'Climate Change and Human Health – Risks and Responses', summary available at http://www.who.int/globalchange/summary/en/index7.html

p. 224   in June 1974 they published their findings in *Nature*: M. Molina and S. Rowland, 'Stratospheric Sink for Chlorofluoromethanes: Chlorine Atom-Catalysed Destruction of Ozone', *Nature*, vol. 249, p. 810 (1974).

p. 225   When the US National Academy of Sciences issued its report: 'Halocarbons, Effects on Stratospheric Ozone', National Research Council (U.S.) Panel on Atmospheric Chemistry, available through Google books at http://books.google.com/books?id=a2YrAAAAYAAJ&dq=Halocarbons:+Effects+on+Stratospheric+Ozone

p. 226   written by a microbiologist in the pages of *Nature*: T. Jukes, 'Spray No More, Ladies', *Nature*, vol. 257, p. 441 (1975). The sardonic tone of this piece makes strange reading today. Jukes, a regular columnist in *Nature* at the time, talks of physicists using the ozone scare to 'cover themselves in glory – their abstruse field is Being Put To Use, while visions of research grants dance through their heads'. He contends that 'the task of the courts in denying appeals against a ban is made childishly simple by the magic word "cancer"', and that the environment is full of checks and balances and will almost certainly dissipate the threat.

p. 229   Perhaps that's why McPeters has since claimed to be the first to report the ozone hole: F. Pukelsheim, 'Robustness of Statistical Gossip and the Antarctic Ozone Hole', *Institute of Mathematical Statistics Bulletin*, vol. 19, p. 540 (1990).

p. 229   His team sent their findings to *Nature* in December: J. Farman, 'Large Losses of Total Ozone in Antarctica Reveal Seasonal ClO_x/NO_x Interaction', *Nature*, vol. 315, p. 207 (1985).

p. 229   According to the historian of science Maureen Christie: M. Christie, 'Data Collection and the Ozone Hole: Too Much of a Good Thing?', *Proceedings of the International Commission on History of Meteorology*, vol. 1.1, p. 99 (2004), available at www.meteohistory. org/2004proceedings1.1/pdfs/11christie.pdf

p. 229   Experts at the United Nations Environment Programme: 'New Report Highlights Two-Way Link Between Ozone Layer and Climate Change', 16 September 2010, available at http://www.unep.org/ Documents.Multilingual/Default.asp?DocumentID=647&Articl eID=6751&1 =en&t=long

p. 230   In 1963, Dennis Gabor published a book: D. Gabor, *Inventing the Future* (Secker & Warburg, 1963).

p. 231   In a 1963 CBS documentary: *CBS Reports: The Silent Spring of Rachel Carson*, first aired on 3 April 1963. You can see the clips in the documentary on Rachel Carson, *The American Experience: Rachel Carson's Silent Spring*, at http://www.youtube.com/watch?v=K-NAUkyIg-M. This is the source for many quotes in this section.

p. 231   Carson put across the extent of the threat with poetic clarity: R. Carson, *Silent Spring* (Houghton Mifflin, 1962).

p. 232   Emil Mrak, a food scientist and the chancellor of the University of California at Davis: Z. Wang, 'Responding to *Silent Spring*: Scientists, Popular Science Communication, and Environmental Policy in the Kennedy Years', *Science Communication*, vol. 19, p. 141 (1997), quoted in N. Oreskes and E. Conway, *Merchants of Doubt*.

p. 232   Carson has been called the 'fountainhead of the modern environmental movement': L. Lear, *Rachel Carson: Witness for Nature* (Mariner Books, 1999), p. 464.

p. 233   on the lawn of the Newagen Inn: You can read the letter at the Inn's website: http://www.boothbayharborblog.com/newagen_seaside_ inn/2008/08/rachel-carsons-last-letter-from-newagen-inn---but-most-of-all-i-shall-remember-the-monarchs.html

p. 234   Hansen took his grandchildren out into the wilds of eastern Pennsylvania: He tells the story in J. Hansen, *Storms of My Grandchildren* (Bloomsbury, 2009), p. 271.

p. 235   In 1988, the US Congress asked Hansen for an opinion: Hansen's response is available at http://image.guardian.co.uk/sys-files/ Environment/documents/2008/06/23/ClimateChangeHearing1988. pdf

p. 236   Joseph Romm summed up the problem: J. Romm, '*Scientific American* Jumps the Shark', 26 October 2006, http://climateprogress.org/2010/10/26/ scientific-american-jumps-the-shark-online-polls-judith-curry/

p. 237   'It seems to me that scientists downplaying the dangers ...': J. Hansen, 'Huge Sea Level Rises Are Coming – Unless We Act Now', *New Scientist*, 25 July 2007, p. 30.

p. 237   Hansen's response to the conservatism of the IPCC: J. Hansen, 'Scientific Reticence and Sea Level Rise', *Environmental Research Letters*, vol. 2, p. 024002 (2007).

p. 237   NASA's attempt to silence Hansen: A.C. Revkin, 'Climate Expert Says NASA Tried to Silence Him', *New York Times*, 29 January 2006, available at http://select.nytimes.com/gst/abstract.html?res=F30D13 FF355B0C7A8EDDA80894DE404482&fta=y&incamp=archive:arti cle_related

p. 238   The organisers celebrated the event: You can read their report at http://capitolclimateaction.org/2009/03/02/ victory-this-is-how-to-stop-global-warming/

p. 238   These days he is advocating putting legal pressure on governments: J. Hansen, 'Activist', in J. Fair, *The Day After Tomorrow: Images of Our Earth in Crisis* (PowerHouse Books, 2010), p. 78. The text is available at http://www.columbia.edu/~jeh1/mailings/2010/20100824_Activist. pdf

p. 240   In a hearing about ozone depletion held before the US Senate: L. Dotto and H. Schiff, *The Ozone War*, p. 194.

p. 240   Susan Solomon took the same stance: A.C. Revkin, 'The Road from Climate Science to Climate Advocacy', *Dot Earth*, 9 January 2008, available at http://dotearth.blogs.nytimes.com/2008/01/09/ the-road-from-climate-science-to-climate-advocacy/

p. 241   a special responsibility to engage in activism: M.C. Nisbet, 'Do Scientists Have a Special Responsibility to Engage in Political Advocacy?' Age of Engagement at Bigthink.com, 13 September 2010, available at http://bigthink.com/ideas/24012. See also J.A. Vucetich and M.P. Nelson, 'The Moral Obligations of Scientists', *Chronicle of Higher Education*, 1 August 2010, available at www.fw.msu.edu/documents/MoralObligationsOfScientists.pdf

p. 241   Carl Sagan put it thus: C. Sagan, *The Demon-Haunted World* (Ballantine, 1996), p. 291.

EPILOGUE

p. 243   Ten days after Crick's death in 2004: A. Rees, 'Nobel Prize Genius Crick Was High on LSD When He Discovered the Secret of Life', *Mail on Sunday*, 8 August 2004 (and personal conversations with Rees).

p. 243   In 1967 he signed a letter to the London *Times*: 'The Law Against Marijuana Is Immoral in Principle and Unworkable in Practice', *The Times* (advertisement), 27 July 1967, p. 5.

p. 243   the idea that Crick used LSD to open his mind: M. Ridley, *Francis Crick: Discoverer of the Genetic Code* (Harper Perennial, 2008), p. 156.

p. 244   Sandoz made it 'readily available …': D. Nichols, 'LSD: Cultural Revolution and Medical Advances', *Chemistry World*, January 2006, available at http://www.rsc.org/chemistryworld/Issues/2006/January/LSD.asp

p. 244   she had to stab the stiletto heel of her shoe into his foot: Personal communication; the source wishes to remain anonymous.

p. 244   Crick regularly smoked pot and used LSD later in life: Ridley, *Francis Crick*, p. 156.

p. 244   'He told me about a lot of private things …': Cristof Koch, personal email communication.

p. 245   'cheer up and take it from us that even if we kicked you in the pants …': A. Gann and J. Witowski, 'The Lost Correspondence of Francis Crick', *Nature*, vol. 467, p. 519 (2010).

p. 245   She was 'too determined to be scientifically sound and to avoid shortcuts': F. Crick, 'How to Live with a Golden Helix', *The Sciences*, September 1979, p. 6, available at profiles.nlm.nih.gov/SC/B/C/D/V/_/scbcdv.pdf

p. 245   'Rosalind, it seems to me, was too cautious.': F. Crick, Letter to Charlotte Friend, 18 September 1979, held in the Wellcome Library for the History and Understanding of Medicine, available at http://profiles.nlm.nih.gov/SC/B/B/X/T/

p. 246   Richard Feynman, enjoyed marijuana and LSD: This is mentioned in J. Gleick, *Genius: The Life and Science of Richard Feynman* (Vintage, 1993), p. 406.

p. 246   Carl Sagan was also a regular user of cannabis: Sagan describes his experiences in an anonymous essay for publication in L. Grinspoon, *Marihuana Reconsidered* (Harvard University Press, 1971), p. 109. 'Mr X' is available at http://hermiene.net/essays-trans/mr_x.html

p. 248   his co-recipients of the 1962 Nobel Prize, were furious: M. Ridley, 'Neither Fish nor Flesh', *The Author*, Summer 2010, p. 54.

p. 248   'It is a layman's illusion': P. Medawar, *The Limits of Science* (Oxford University Press, 1984), p. 101.

p. 249   'In terms of the fulfilment of declared intentions …': P. Medawar, *The Limits of Science*, p. 65.

p. 249   each time a natural limit has been suggested, it has been exceeded: R. Smith, 'A Brief History of Ageing', *Spotlight*, September 2007, available at http://www.research-horizons.cam.ac.uk/spotlight/a-brief-history-of-ageing.aspx

p. 250   he railed against having to modify a paper to meet the objections of his colleagues: D. Kennefick, 'Einstein Versus the *Physical Review*', *Physics Today*, September 2005, p. 43. See also 'Three Myths About Scientific Peer Review', 8 January 2009, a fascinating blog post by Michael Nielsen at http://michaelnielsen.org/blog/three-myths-about-scientific-peer-review/

p. 250   The famous Crick and Watson paper on the structure of DNA was also not peer-reviewed: J. Maddox, 'How Genius Can Smooth the Road to Publication', *Nature*, vol. 426, p. 119 (2003).

p. 252   '… not the only way to ensure quality control in science': M. Rees, 'Not Worth the Paper', *New Scientist*, 23 November 2002, p. 27.

p. 253   'Scientists should be on tap, but not on top': W. Churchill, *Twenty-One Years* (Weidenfeld & Nicolson, 1964), p. 127.

p. 254   'It is a perversion of democracy to muffle the voice of the most knowledgeable …': M. Nelson and J. Vutevich, 'The Moral Obligations of Scientists', *Chronicle of Higher Education*, 1 August 2010, available at www.fw.msu.edu/documents/MoralObligationsOfScientists.pdf

p. 254   A study in Scotland, for instance, approached nineteen committees: K. Ah-See et al., 'Local Research Ethics Committee Approval for a National Study in Scotland', *Journal of the Royal College of Surgeons of Edinburgh*, vol. 43, p. 303 (1998).

p. 255   'The burdens imposed by ethics review might be justified ...': P. Glasziou and I. Chalmers, 'Ethics Review Roulette: What Can We Learn?', *British Medical Journal*, vol. 328, p. 121 (2004).

p. 256   As the philosopher Rousseau suggested: J.-J. Rousseau, *Émile* (The Echo Library, 2007), p. 128. Rousseau believed that only when the child is more mature should they be taught methods of learning science.

p. 256   Traditional forms of education, he says, can hamper scientific creativity in students: M. Villiger, 'Meet the Ovshinskys', *Scientific American Frontiers*, 19 May 2004, available at http://www.pbs.org/saf/1506/features/ovshinsky2.htm

p. 257   '... in it for the money because it was suddenly available': K. Mullis, 'What Scientists Do', TED talk, February 2002, available at http://www.ted.com/index.php/talks/kary_mullis_on_what_scientists_do.html

p. 257   a 2008 paper by the GEM particle physics collaboration: A Budzanowski et al., 'Cross Section and Tensor Analysing Power of the $dd \to \eta\alpha$ Reaction Near Threshold', *Nuclear Physics A*, vol. 821, p. 193 (2009). If you really want to, you can download the paper from http://arxiv.org/abs/0811.3372v1

p. 257   the majority of scientists are 'shut up in the narrow cell of their laboratory ...': O. Gasset, *The Revolt of the Masses* (Unwin, 1969), p. 85.

p. 258   'Never lose sight of the role your particular subject has ...': E. Schrödinger, *Science and Humanism* (Cambridge University Press, 1951), p. 8.

p. 258   'If you follow the herd, all the grass is gone': Andre Geim, interview with the author for *New Scientist*, 14 October 2010.

p. 260   'These are the marks of science ...': J. Bronowski, *The Common Sense of Science* (Heinemann, 1951), p. 150.

p. 260   those who have felt the 'torment of the unknown.': C. Bernard, *An Introduction to the Study of Experimental Medicine* (Dover Publications, 1927), p. 222.

p. 260   'Science is the acceptance of what works and the rejection of what does not': J. Bronowski, *The Common Sense of Science*, p. 148.

INDEX